Extensive Aquaculture

Extensive Aquaculture

Bruno Augusto Amato Borges

Delve
PUBLISHING
www.delvepublishing.com

Extensive Aquaculture

Bruno Augusto Amato Borges

Delve Publishing
224 Shoreacres Road
Burlington, ON L7L 2H2
Canada
www.delvepublishing.com
Email: orders@arclereducation.com

ISBN: 978-1-77469-129-8 (Hardcover)

ABOUT THE AUTHOR

Bruno Augusto is Aquaculture Engineer and Master on Aquaculture and Fishery Recources from Federal University of Santa Catarina (UFSC). His focus is on writing projects and development of new technologies in aquaculture. He is currently project service provider, consultant and develops research on Biofloc Technology (BFT) cultivation for fish and shrimp.

TABLE OF CONTENTS

LIST OF FIGURES

ABSTRACT

The aquaculture sector has a lot to add towards achieving the SDGs, especially SDG 14 –which involves conserving and economically utilizing the seas, oceans and marine resources for sustainability. Aquaculture keeps on showing huge success in providing food, sustenance and work. This book shows the significant challenges ahead in spite of the advancement made on various fronts. For instance, there is developing proof that when fisheries are appropriately overseen, stocks are reliably above target levels, offering validity to the fishery stakeholders and governments all throughout the planet. Notwithstanding, the book likewise shows that the victories accomplished in certain nations have not been adequate to invert the global pattern of overfished stocks, demonstrating that in places where fishery boards aren't set up, or are insufficient, the fish stocks become poor and weakened. This calls for new components to help in the viable execution of strategy and the guidelines for sustainable fisheries and environments, as there is no lone answer for guaranteeing fisheries stock throughout the planet.

PREFACE

Aquaculture has developed quickly over the years, and today it creates nearly as much fish and shellfish as mainstream fisheries. It is the fundamental means for acquiring more food from our aquatic surroundings. Effects of aquaculture on biodiversity emerge from the utilization of resources, like land (or srate), water, seed, and feed, their change into items valueed by society, and the resulting discharge into the climate of ozone-depleting substances, microorganisms and parasites. Adverse consequences might be immediate, through the arrival of eutrophicating substances, poisonous synthetic compounds, the exchange of microorganism disease and parasites to wild population stock, and the presentation of hereditary material into the environment, or circuitous through loss of environment and niche srates and changes in food networks.

Today, huge amounts of fish are gotten from aquaculture– which other than being a good use of food resources likewise may add to reducing the high overfishing on certain stocks with expected unfortunate results for marine food networks. Regardless of advances being made in the feed business, bringing about diminished feed transformation proportions and improvement of appropriate choices to fish resources, the book specifies the aquaculture business' utilization of worldwide fishmeal and fish oil expanded three-fold somewhere between 1992 and 2006.

The existence cycles for most aquatic species have been effectively shut yet the way of life of a few, particularly marine, cultivated fish, and shrimp, is still part of the way reliant upon the catch of hatchlings, postlarvae, or gravid females from nature. This can bring about both over-fishing and bycatch, causing losses to catch fisheries and biodiversity. Enormous srates of natural environments, for example, wetlands and mangroves have generally been lost because of aquaculture improvements and contamination, bringing about diminished biodiversity and enlistment to catch fisheries. Today the upsides of seaside wetlands are better perceived and guidelines make it hard for large-scale aquaculture advancement in beach front regions (e.g., mangroves, seagrass beds) in many nations.

The size of biodiversity shortfall from aquaculture improvement for the most part increases with scale, the force of resource use, and net production of wastes. It is additionally a lot of work on the area of production site, species being refined, the aquatic framework being utilized, strategy for development and the directors, and the aquaculture venue. The book proposes sometimes that aquaculture may likewise expand nearby biodiversity, for instance, when lakes are developed in dry regions, through re-loading exercises and from integrated aquaculture. Its contribution in keeping up with social diversity should likewise be recognized, aquaculture creates employment and boosts the socioeconomic condition of people.

This book focuses on nations where the foundation, public limits and global help are generally restricted, and where effective coordinated effort and associations to foster aquaculture skills and innovation can be of specific advantage. For instance, environmental change impacts on marine catch fisheries are projected to be more prevalent in tropical areas of Africa and Asia, where warming diminishes usefulness. Designated fisheries and aquaculture improvement mediations in these districts, tending to their particular requirements for food, and jobs, can give the groundbreaking needed to take care of everybody, all over. Aquatic food networks influence the climate, yet that there are compromises to guarantee we further develop food security while limiting the effects on their steady biological systems.

Chapter 1

Global Demand for Dietary Fish Protein

CONTENTS

1.1 INTRODUCTION—FISH ARE A RESOURCE FOR PEOPLE

Fish give nourishment, occupations, and prosperity to individuals. We eat them for wellbeing and custom; we swim with them for otherworldliness and fun; we chase them for game and entertainment; we go to aquariums to watch them with family; we depend on them for local area and work. The United Nations estimates that fish support the occupations of 10-12% of the world.

Fish assumes a significant part in worldwide maintainability and protection, however isn't regularly found in a worldwide, ecological setting. Seeing how fish production and consumption finds a way into worldwide food systems and protection is significant for understanding why fish is regularly the best animal protein to eat to battle deforestation, water shortage, and environmental change—the biggest danger mankind faces (Verdegem & Bosma, 2009).

What does the world eat?

Your body needs calories with the right equilibrium of fats, starches, proteins, and micronutrients (nutrients and minerals). The measure of food the body needs is dictated by an individual's weight and digestion, however dietary rules set by the U.S. government proposed getting 45-65% of calories from carbs; 20-35% from fats, and 10-35% from proteins. Food varieties are sorted by their predominant supplement Flarge e.g., bread contains protein, yet is for the most part grain so consequently viewed as a starch); fish is a protein, explicitly an animal protein.

Proteins can be utilized for energy, yet their fundamental job is to give a construction to cells in the body—particularly muscle and tissue cells. Proteins are made out of amino acids, of which there are 9 various types considered "essential amino acids" that our bodies need to work appropriately. Food proteins that contain every one of the 9 of these amino acids is classified as "complete proteins". FAO expresses that individuals burning through 75% of their weight (in kg) in grams of protein each day are protein adequate (for instance, on the off chance that somebody weighs 80kg, they ought to burn through 60g of protein each day).

Figure 1.1: Common proteins.

Source: https://sustainablefisheries-uw.org/seafood-101/what-does-the-world-eat/

People as omnivores

All meat, from pork to poultry to hamburger, and fish, are complete proteins with a high, adjusted measure of the 9 fundamental amino acids. Plant-based proteins are not thought of as "complete" because practically all plant proteins need at least one fundamental amino acids. Plant-based proteins are additionally regularly imbalanced—high in certain sorts of amino acids, yet low in others. It is not necessarily the case that veggie lovers and vegetarians are not getting the entirety of their fundamental amino acids—consuming the right assortment of plant-based proteins can supply the body with each of the 9 fundamental ones. In any case, it tends to be hard to manage without admittance to a scope of food varieties (Van Doan et al., 2020). Numerous individuals throughout the planet are restricted to food sources delivered in their nearby environments; a reliable assortment of food requires stable systems of exchange, transport, and refrigeration—a test in a large part of the world. Indeed, even the individuals who live in nations with numerous food imports can in any case battle with access. For instance, in the USA, more than 25 million individuals live in a "food desert" without helpful admittance to a supermarket or grocery store. For some individuals, it is simplest to get all fundamental amino acids by eating meat.

Be that as it may, not all meat is similarly sound; the fat substance decides the number of calories a serving contains. High-fat meats, similar to hamburgers, contain undeniably more calories for a similar measure of protein than less fatty meats like poultry or fish.

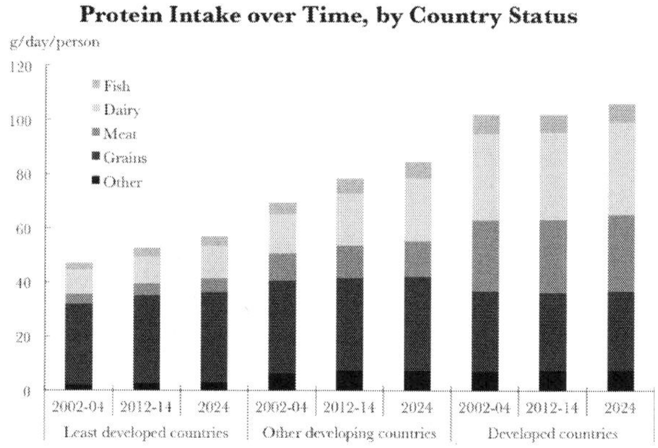

Figure 1.2: Protein intake over time by country status.

Source: https://sustainablefisheries-uw.org/seafood-101/what-does-the-world-eat/

With the advantages of abundance and exchange, developed countries consume substantially more protein, particularly complete proteins, than more undeveloped countries. As nations create and secure resources and exchange, interest for meat rises, sustenance improves, and by and large human wellbeing improves.

1.2 FISH AS PROTEIN

However fish make a limited quantity out of worldwide protein requirement (6.7%), they are a significant source of animal protein, giving 17% of the world's meat utilization. What's more, fish assume an outsized dietary part for some individuals; 3.1 billion individuals depend on fish for 20% of their everyday protein consumption, for certain coastal networks are dependent on fish for as much as 70%.

As a significant source of supplements, worldwide fish furnishes around three billion individuals with very nearly 20% of their normal per capita requirement of animal protein, and 4.3 billion individuals with around 15% of such protein, as per the FAO's "The State of World Fisheries and Aquaculture 2012" report. Fish and fishery items address a significant source of supplements of crucial significance for expanded and solid weight control plans. With a couple of exemptions for chose species, fish is generally low in immersed fats, carbs and cholesterol. Fish offers high-benefit protein yet, in addition, a wide scope of fundamental micronutrients, including different vitamins (D, A and B), minerals (counting calcium, iodine, zinc, iron and selenium) and polyunsaturated omega-3 unsaturated fats (docosahexaenoic acid and eicosapentaenoic acid). While normal per capita fish utilization might be low, even little amounts of fish can have a critical positive healthful effect by giving fundamental amino acids, fats and micronutrients that are scant in vegetable-based weight control plans. There is proof of advantageous impacts of fish utilization comparable to coronary illness, stroke, age-related macular degeneration and psychological wellness. There is additionally persuading proof regarding benefits as far as development and improvement, specifically for ladies and youngsters during growth and outset for the ideal mental health of kids (Turcios & Papenbrock, 2014).

By and large, fish gives just around 33 calories for each capita each day. Be that as it may, it can surpass 150 calories for every capita each day in nations where there is an absence of elective protein food and where an inclination for fish has been created and kept up with (for example Iceland, Japan and a few little island States). The dietary commitment of fish is more critical as far as animal proteins, as 150 g of fish gives around 50-60% of the daily protein prerequisites for a grown-up. Fish proteins can address an essential part in some thickly populated nations where complete protein consumption levels might be low. Indeed, numerous populations, more those in agricultural nations than created ones, rely upon fish as a component of their everyday diet. For them, fish and fishery items frequently address a reasonable source of animal protein that may not exclusively be less expensive than other animal protein sources, yet liked and part of the neighborhood and conventional plans. For instance, fish adds to or surpasses, 50% of all-out animal protein consumption in some little island creating States, just as in Bangladesh, Cambodia, Ghana, the Gambia, Indonesia, Sierra Leone and Sri Lanka. In 2009, fish represented 16.6 percent of the worldwide requirement of animal protein and 6.5 percent of all protein consumed through. All around the world, fish furnishes about 3.0 billion individuals with very nearly 20% of their normal per capita requirement of animal protein, and 4.3 billion individuals with around 15% of such protein.

Connected to the solid development of fish production and present-day circulation channels, the world fish food supply developed at a normal rate of 3.2 percent each year in the period 1961-2009, dominating the increment of 1.7 percent each year on the planet's population; thus, normal per capita accessibility has risen. World per capita fish utilization expanded from a normal of 9.9 kg during the 1960s to 11.5 kg during the 1970s, 12.6 kg during the 1980s, 14.4 kg during the 1990s, 17.0 kg during the 2000s and arrived at 18.4 kg in 2009.

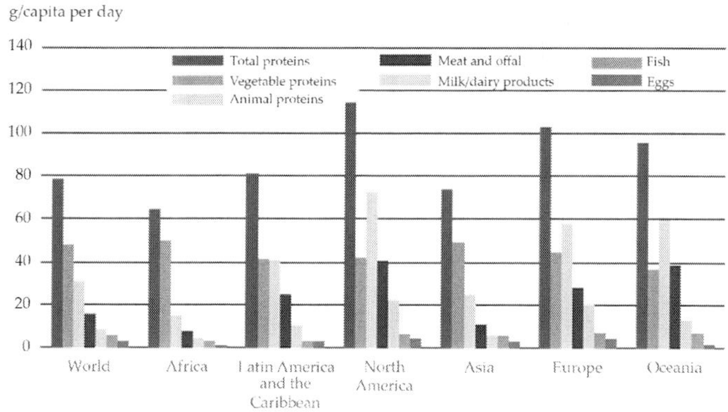

Figure 1.3: Total protein supply by continent and major food group.

Source: https://thefishsite.com/articles/fao-state-of-world-fisheries-aquaculture-report-fish-consumption

Primer assessments for 2010 point towards a further expansion in per capita fish utilization to 18.6 kg. It ought to be noticed that figures for 2000 are higher than those revealed in past releases of The State of World Fisheries and Aquaculture, as FAO has updated downwards the non-food estimates for China consumption, beginning from 2000 information, to think about further developed public data the area. As an outcome, per capita, fish utilization figures for China just as for the world have expanded in examination with past appraisals. Despite the solid expansion in the accessibility of fish to most purchasers, the development in fish consumption contrasts extensively among nations and inside nations and districts as far as amount and assortment consumed per head. For instance, per capita, fish consumption has stayed static or diminished in certain nations in sub-Saharan Africa (for example the Congo, South Africa, Gabon, Malawi and Liberia) and Japan over the most recent twenty years, while the most considerable expansions in yearly per capita fish consumption have happened in East Asia (from 10.6 kg in 1961 to 34.5 kg in 2009), Southeast Asia (from 12.8 kg in 1961 to 32.0 kg in 2009) and North Africa (from 2.8 kg in 1961 to 10.6 kg in 2009). China has been answerable for the majority of the increment in world per capita fish consumption, attributable to the considerable expansion in its fish production, specifically from aquaculture (Troell et al., 2014). China's share in world fish production developed from 7% in 1961 to 34 percent in 2009. Per capita fish production in China has likewise expanded drastically, coming to about 31.9 kg in 2009, with a normal yearly development rate of 4.3 percent in the period 1961-2009 and of 6.0 percent in the period 1990-2009. Over the most recent couple of years, fuelled by developing homegrown pay and resources, customers in China have encountered an expansion of the kinds of fish accessible inferable from a redirection of some fishery sends out towards the homegrown market just as an increment in fishery imports. In case China is rejected, yearly per capita fish supply to the remainder of the world was about 15.4 kg in 2009, higher than the normal upsides of the 1960s (11.5 kg), 1970s (13.5 kg), 1980s (14.1 kg) and 1990s (13.5 kg). It ought to be noticed that during the 1990s, world per capita fish supply, barring China, was somewhat steady at 13.113 kg and lower than during the 1980s as the population developed more quickly than food fish supply (at yearly rates of 1.6 and 0.9 percent, separately). Since the mid-2000s, there has been a reversal of this pattern, with food fish supply development dominating population development (at yearly rates of 2.6 percent and 1.6 percent, separately).

	Total food supply	**Per capita food supply**
	(million tonnes live weight equivalent)	**(kg/year)**
World	125.6	18.4
World (excluding China)	83.0	15.1
Africa	9.1	9.1
North America	8.2	24.1
Latin America and the Caribbean	5.7	9.9
Asia	85.4	20.7
Europe	16.2	22.0
Oceania	0.9	24.6
Industrialized countries	27.6	28.7
Other developed countries	5.5	13.5
Least-developed countries	9.0	11.1
Other developing countries	83.5	18.0
LIFDCs[2]	28.3	10.1

[1]Preliminary data.

[2]Low-income food-deficit countries.

Figure 1.4: Total and per capita food fish supply by continent and economic grouping in 2009.

Source: https://thefishsite.com/articles/fao-state-of-world-fisheries-aquaculture-report-fish-consumption

The table above sums up per capita fish consumption by mainland and major monetary gathering. The aggregate sum of fish consumed and the specific arrangement of the food fish supply change as per districts and nations, mirroring the various degrees of accessibility of fish and different food sources, remembering the availability of fishery assets for neighboring waters just as the cooperation of a few financial and social components. These elements incorporate food customs, tastes, requests, pay levels, seasons, costs, wellbeing framework and correspondence offices. Yearly per capita clear fish consumption can fluctuate from under 1 kg in one nation to more than 100 kg in another. Contrasts may likewise be critical inside nations, with consumption typically higher in seaside, riverine and inland water regions. Of the 126 million tons accessible for human consumption in 2009, fish consumption was least in Africa (9.1 million tons, with 9.1 kg per capita), while Asia represented 66% of all-out consumption, with 85.4 million tons (20.7 kg per capita), of which 42.8 million tons was consumed outside China (15.4 kg per capita). The comparing per capita fish consumption figures for Oceania, North America, Europe, and Latin America and the Caribbean were 24.6 kg, 24.1 kg, 22.0 kg and 9.9 kg, individually.

Contrasts in fish consumption exist between the more-created and the less developed nations. Albeit yearly per capita consumption of fishery items has filled consistently in creating areas (from 5.2 kg in 1961 to 17.0 kg in 2009) and in LIFDCs (from 4.9 kg in 1961 to 10.1 kg in 2009), it is still extensively lower than that of more created districts, even though the gap is narrowing. The genuine qualities might be higher than shown by true insights taking into account the under-recorded commitment of means fisheries and some limited scale fisheries. In 2009, clear per capita fish consumption in industrialized nations was 28.7 kg, while for all developed countries it was assessed at 24.2 kg. A sizeable portion of fish burned in developed countries comprises of imports, and attributable to consistent interest and declining homegrown fishery production (down 10% in the period 2000 to 2010), their reliance on imports, specifically from non-industrial nations, is projected to develop. In agricultural nations, fish consumption will in general be founded on locally and occasionally accessible items, and the fish chain is driven by supply instead of interest. Notwithstanding, in arising economies, imports of fishery items not accessible locally have as of late been developing (Teletchea & Fontaine, 2014).

Incongruities among created and non-industrial nations exist likewise concerning the commitment of fish to animal protein usage. Despite their generally low degrees of fish consumption, this offer was huge at about 19.2 percent for agricultural nations and 24.0 percent for LIFDCs. Notwithstanding, this offer has declined somewhat lately inferable from the developing consumption of other animal proteins. In developed nations, the portion of fish in animal protein requirement, after steady development up to 1989, declined from 13.9 percent in 1984 to 12.4 percent in 2009, while consumption of other animal proteins kept on expanding.

The fish area stays exceptionally divided, specifically for business sectors of new fish, yet it is in a period of combination and globalization. Fish is exceptionally heterogeneous and contrasts might be founded on species, production region, strategy for fishing or cultivating, taking care of training and cleanliness. Crude fish can be handled into a much more extensive scope of items to fulfill shopper needs that vary as indicated by business sectors, adaptability in supply volumes, actual closeness, providers reliability, capacity to adjust to various segment size particulars, and so on. Over the most recent twenty years, the consumption of fish and fishery items has additionally been impacted significantly by globalization in food systems and by developments and enhancements in handling, transportation, circulation, promoting and food science and innovation. These variables have prompted critical upgrades in proficiency, lower costs, more extensive decisions and more secure and further developed items. Inferable from the perishability of fish, advancements in long-distance refrigerated vehicles and huge scope and quicker shipments have worked with the exchange and consumption of an extended assortment of species and item shapes, including live and new fish. Consumers can profit with an expanded decision, with imports boosting the accessibility of fish and fishery items in the homegrown business sectors.

Developing interest from nearby consumers has likewise supported aquaculture advancement in numerous locales in Asia and, progressively, in Africa and Latin America. Since the mid-1980s, and specifically, somewhat recently, the commitment of aquaculture to fish consumption has shown sensational development, as catch fisheries production has deteriorated or even declined in certain nations. In 2010, aquaculture contributed around 47% of the fishery yield for human consumption amazing development contrasted and its 5% in 1960, 9 percent in 1980 and 34 percent in 2000, with a normal yearly development rate of 4.7 percent in the period 1990-2010. Nonetheless, in case China is rejected, the normal target of aquaculture is essentially lower at 17% in 2000 and 29 percent in 2010, relating to a normal yearly development rate of 5.4 percent (Sirakov et al., 2015). Aquaculture has pushed the interest for, and consumption of, species that have moved from being principally wild-caughtto being aquaculture created, with a reduction in their costs and a solid expansion in their commercialization, for example, for shrimps, salmon, bivalves, tilapia, catfish and Pangasius. Aquaculture additionally assumes a part in food security through the critical production of some low-value freshwater species, which are principally bound for homegrown production, likewise through incorporated cultivating.

The flooding commitment of aquaculture can likewise be noted by noticing fish consumption by significant gatherings. Attributable to the expanding production of shrimps, prawns and mollusks from aquaculture and the general decrease in their value, yearly per capita accessibility of shellfish developed considerably from 0.4 kg in 1961 to 1.7 kg in 2009, and that of mollusks (counting cephalopods) rose from 0.8 kg to 2.8 kg in a similar period. The expanding production of salmon, trouts and chose freshwater species has prompted a critical development in a yearly per capita consumption of freshwater and diadromous species, up from 1.5 kg in 1961 to 6.0 kg in 2009. Over the most recent couple of years, no significant changes have been capable by the other more extensive gatherings. Yearly consumption of demersal and pelagic fish species has settled at about 3.0 kg and 3.4 kg per capita, separately. Demersal fish keep on being among the primary species supported by consumers in Northern Europe and North America (8.6 kg and 7.0 kg per capita each year, individually, in 2009), while cephalopods are principally liked by the Mediterranean and East Asian nations. Of the 18.4 kg of fish per capita accessible for consumption in 2009, around 74 percent came from finfish. Shellfish provided 26% (or about 4.5 kg per capita, partitioned into 1.7 kg of crustaceans 0.5 kg of cephalopods and 2.3 kg of different mollusks).

The worldwide development in fish utilization mirrors patterns in food utilization overall. Per capita, food utilization has additionally been filling over the most recent couple of many years. Except for the times of the food and financial emergencies, the worldwide food market, including the fish market, has encountered uncommon development and an adjustment of worldwide dietary examples, turning out to be more homogeneous and globalized. This change has been the consequence of a few variables, including rising expectations for everyday comforts, population development, fast urbanization and openings for exchange and changes in food dispersion. A blend of these variables has prompted developing interest in protein food items, specifically meat, fish, milk, eggs just as vegetables in the eating routine, with a decrease in the portion of staples like roots and tubers. Protein accessibility has expanded in both the developed and developing world, yet this development has not been similarly disseminated. There has been an amazing expansion in the utilization of animal items in nations like Brazil and China and other less-created nations. As per FAOSTATS, yearly worldwide per capita meat utilization developed from 26.3 kg in 1967 to 32.4 kg in 1987 to arrive at 40.1 kg in 2007. The development was specially set apart in the most quickly arising economies of agricultural nations, and yearly per capita meat utilization in non-industrial nations dramatically increased from 11.2 kg in 1967 to 29.1 kg in 2007 (Stenton-Dozey et al., 2021). The stock of animal protein keeps on leftover altogether higher in industrialized and other developed countries than in non-industrial nations. Notwithstanding, having accomplished a significant degree of utilization of animal protein, more created economies have been progressively arriving at immersion levels and are less receptive than low-pay nations to pay development and different changes. In developed nations, per capita, meat utilization expanded from 61.4 kg in 1967 to 80.7 kg in 1987, yet then, at that point declined to 75.1 kg in 1997 preceding arriving at 82.9 kg in 2007.

Despite the improvement in per capita accessibility of food and the positive long haul patterns in healthful guidelines, undernutrition (counting lacking degrees of utilization of protein-rich food of animal beginning) stays a colossal and determining issue. Lack of healthy sustenance is a significant issue around the world, with one individual in seven undernourished and more than 33% of infant mortality owing to undernutrition. This is particularly the situation in many agricultural nations, with the heft of undernourished individuals living in country regions.

The food area overall is experiencing a time of underlying change because of developing wages, alterations in the population structure, new ways of life, globalization, progression of exchange and the rise of new business sectors. A more prominent center is additionally being given to advertising, with producers and retailers turning out to be more mindful of shopper inclinations and endeavoring to expect market assumptions as far as quality, wellbeing principles, assortment, value option, and so on. Consumer propensities have changed essentially in the previous few decades, and food issues like a guilty pleasure, accommodation, wellbeing, morals, assortment, an incentive for cash, and security are turning out to be progressively significant, particularly in more well-off economies.

In these business sectors, customers are mentioning better expectations as far as food newness, variety, comfort and wellbeing, including quality affirmations like detectability, pressing necessities and preparing controls. Purchasers presently request that their food has been created, dealt with and sold in manners that protect their wellbeing, regard the climate and address different moral and social concerns. Well-being and prosperity are progressively impacting utilization choices, and fish has a specific noticeable quality in this regard, following mounting proof affirming the medical advantages of eating fish. This is halfway identified with a maturing society, yet food handling issues just as weight and unfavorably susceptible responses have additionally brought issues to light about the connection between food and wellbeing. In more-developed economies, fast decreases in aging joined with upgrades in endurance are prompting a maturing population, wherein an expanding extent of the population is concentrated among more elderly groups. In numerous nations of the more evolved areas, more than 20% of the population is matured 60 or over. This influences the interest in various kinds of food.

These continuous changes in shopper inclinations are progressively affecting mechanical advancements and new systems for getting sorted out the inventory network. Most item advancements manage steady adjustments, like varieties in taste and bundles intended for various types of utilization. World food markets have gotten more adaptable, with new items entering the business sectors, including value added items that are simpler for consumers to get ready (Sorgeloos, 2013). Corporate stores, transnational organizations and general stores are additionally arising as a significant power, especially in non-industrial nations, offering consumers a more extensive decision diminished occasional change inaccessibility and, regularly, more secure food. A few agricultural nations, particularly in Asia and Latin America, have encountered a fast development in the number of grocery stores, which are progressively focusing on lower-and center pay customers just as the higher-pay gatherings.

Developing urbanization is one of the elements altering food utilization designs, with an effect additionally on the interest for fishery items. Individuals living in urban regions will in general commit a higher extent of their pay to food bought and to eat out of the home all the more regularly, and to buy bigger amounts of quick and comfort food sources. Also, expanding urbanization intensifies the tension in nearby regions to fulfill the need for huge, concentrated populations. According to the United Nations Population Division (2011), 52.1 percent (3.6 billion individuals) of the earth's population were living in urban regions. Inconsistencies in the degrees of urbanization endure among nations and areas of the world, with more-developed countries having a urban portion of up to 78 percent, while others remain for the most part rural, specifically, LDCs (with a urban portion of around 29%), Africa (40%) and Asia (45%). Nonetheless, likewise in these last regions, a huge development of the population towards the urban communities is occurring. An extra 294 million and 657 million individuals are required to become urbanized by 2015 and 2020, separately, with the heft of the increment in urban regions expected in Asia and Africa. By 2050, the portions of the urban population will be 58% in Africa and 64 percent in Asia, albeit this will in any case be essentially not exactly the most different mainlands. The country population is relied upon to decrease in each significant region besides in Africa.

The viewpoint for the worldwide food area stays questionable. It is confronting different difficulties identified with the monetary turndowns in chose nations and segment issues, including developing urbanization. The drawn-out figure for food requests stays positive, likewise determined by population development and urbanization. Specifically, interest in fish items is relied upon to keep on ascending in the coming many years. Notwithstanding, future expansions in per capita fish utilization will rely upon the accessibility of fishery items. With catch fisheries production deteriorating, significant expansions in fish food production are conjecture to come from aquaculture. In any case, future interest will be controlled by a perplexing collaboration of a few components and components. The worldwide food areas, including the fishery area, should confront a few difficulties coming from the segment, dietary, environmental and monetary changes, remembering diminished dependence for fossil energy and expanding limitations on other regular assets. Specifically, the future organic market of food wares, including fisheries, will be influenced by population elements and the area and rate of monetary development. Total population development is relied upon to moderate in the following decade, in all locales and landmasses, with non-industrial nations proceeding to encounter the fastest population increments.

The worldwide population is set to increment from around 7 billion out of 2011 to 7.3 billion of every 2015 and 7.7 billion out of 2020 and 9.3 billion out of 2050, with the greater part of the expansion happening in non-industrial nations, as per the medium-variation projections arranged by the United Nations. A lot of this increment is projected to come from the developed nations and it is relied upon to happen in urban regions.

1.3 FISH CONSUMPTION IN SELECTED CASE STUDY COUNTRIES

Notwithstanding the expansive outline presented above, we zeroed in on eight nations to give further and more itemized bits of knowledge. The contextual investigation nations were:

- France and Germany (Europe),
- Peru and the United States of (America),
- China and Indonesia (Asia),
- Senegal and South (Africa).

The decision depended on the accompanying standards: (1) every landmass (except for Oceania) ought to be addressed, (2) created just as agricultural nations ought to be incorporated, (3) fish and fisheries should assume a significant part for those nations. The African nations Senegal and South Africa have the most minimal fish consumption, by and large, beginning with 0.06 million tons (Senegal) and 0.1 million tons (South Africa) expanding to generally 0.3 million tons in the two nations. Peru's fish consumption marginally surpasses African qualities with 0.14 million tons in 1961 and 0.65 million tons in 2011. Interestingly, fish consumption in the USA began a lot higher in 1961 with 2.5 million tons and expanded to 6.8 million tons in 2011. Even though Indonesia's fish consumption was under 1 million tons in the mid-1960s, it has arrived at values like the USA as of late with a limit of 6.9 million tons in 2011. The forerunner in fish consumption is China. It multiplied its fish consumption from 3.4 million tons in 1961 to 6.9 million tons in 1984. From the 1980s until now, China's fish consumption developed at extremely high rates, prompting 46 million tons in 2011. In the examination, the advancement of fish consumption was very moderate in the two European nations Germany and France. Fish consumption levels in Germany remained somewhat steady over the long haul at 0.7 million tons in 1961 and 1.2 million tons in 2011. France encountered a more grounded increment, ascending from 0.7 million tons in 1961 to 2.2 million tons in 2011 (Skerritt et al., 2020).

Contrasts between the contextual investigation nations can again be found in the per capita fish consumption after some time. While South Africa's per capita consumption is like the African normal, Senegalese individuals burn through four to multiple times more fish than individuals in South Africa. Peru and the USA both surpass their landmass' normal. Additionally, these two nations show a comparable improvement in per capita fish consumption over the long run. In any case, Peru shows a high vacillation. This changeability is fundamentally determined by the solid reliance (up to 80%) of the Peruvian fishery on anchoveta (*Engraulis ringens*). For instance, in the mid-1980s, anchoveta stocks west of South America diminished – mostly on account of an El Niño occasion.

This decline was reflected in the Peruvian per capita consumption. China and Indonesia encountered a solid expansion in per capita consumption, like the general Asian turn of events. Per capita consumption in Germany and Europe is somewhat consistent over the long run. Germany lies beneath the European normal.

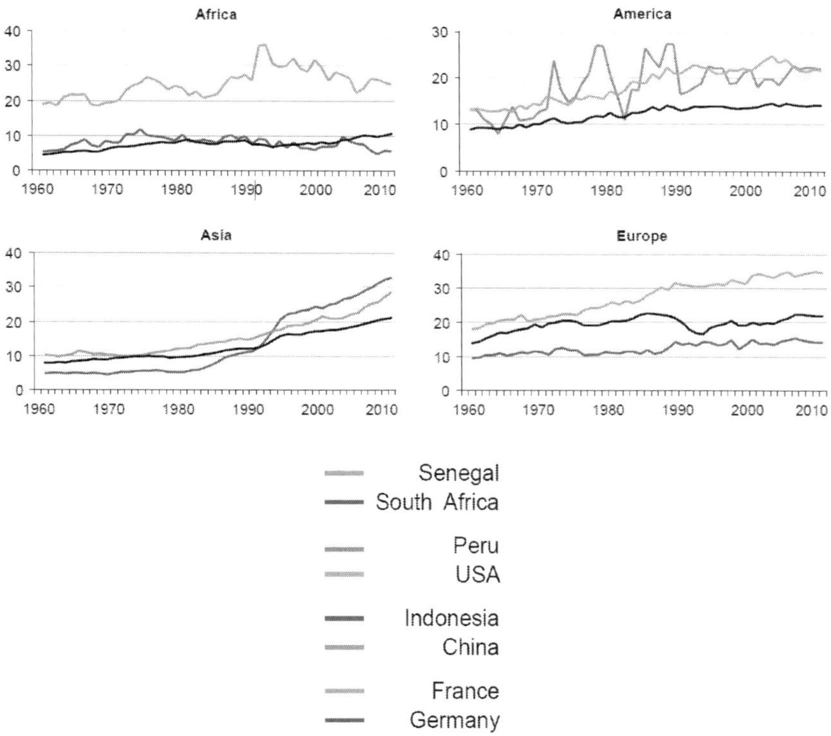

Figure1.5: Per capita fish consumption in the eight case study countries and the corresponding continent (black line) (in kg/year).

Source: https://www.fishforward.eu/wp-content/uploads/2017/11/future_trends_report-EN.pdf

Conversely, France surpasses the European normal as far as fish consumption per individual and encountered a generally solid increment over the long run from 18 kg for each individual to 35 kg for every individual. When contrasting consumption designs at the country level, contrasts between the agricultural nations (South Africa, Senegal, Indonesia and Peru) and the developed countries (China, France, Germany and the USA) can be taken note of. In the agricultural nations, marine fish constitute the greatest portion of eaten fish with pelagic fish unmistakably overwhelming. Additionally, aside from Indonesia, the portion of freshwater fish consumption is exceptionally little. Indonesia is one of the greatest aquaculture producers on the planet. The fundamental freshwater species delivered are carp, tilapia and gourami; shrimp are likewise developed.

A comparative contention may remain constant for Peru, which shows expanded consumption of freshwater species and mollusks from 1990 on. Towards the finish of the 1980s, Peru started the aquaculture production of trout, tilapia, shrimps and scallops that grew effectively in the ensuing years. Among the developed countries considered in this report, Germany and France have the most elevated requirement of marine fish. In any case, over the long run, this offer has diminished in France for mollusks and cephalopods. In Germany, freshwater fish consumption has expanded. In the USA, by and large, consumption has remained moderately steady, while, as in France and Germany, the portion of marine fish has diminished while the portion of freshwater fish has expanded.

The expansion in the consumption of freshwater species, cephalopods contrasted with wild catches. China is the world's driving country in aquaculture production and shows the most noteworthy portion of freshwater fish and mollusks consumption is ongoing for many years. Interestingly, its consumption of pelagic and demersal fish is the littlest of all the eight contextual investigation nations. The quick expansion in fish consumption in the creating economies in Asia can be clarified by the connection between's expanding fish consumption and expanding resources: per capita requirement of fish increments quickest where abundance and urbanization are consolidated and where homegrown inventory is expanding too (Shaalan et al., 2018).

Generally, the portion of marine fish and fish in world consumption has declined over the long run, while the portion of freshwater fish has shown an expansion. Notwithstanding, marine fish frames most of the world's fish consumption, and a few nations, for instance, South Africa, depending predominantly on wild profits. By the by, albeit Sub-Saharan Africa at present offers under 1% of worldwide aquaculture production, it registers as the quickest developing aquaculture industry estimated by its development rate. African aquaculture production is still exceptionally low. There are various explanations behind this, including troublesome economic situations and an emphasis on smallholder aquaculture, which is vital for neighborhood food security yet can't meet the objective of expanded fish production at the public level. The monoculture of fish is progressively supplanting the traditional consumption of little fish species in some low-pay nations.

Little pelagic fish specifically have a one of a kind wholesome arrangement and are subsequently of extraordinary dietary significance. Also, little fish are more moderate and more effectively open than bigger fish or other animal-based food sources and vegetables. Consumption of these ought to be energized and advanced, while the consumption of little pelagic fish for the fish feast and fish oil ought to be reevaluated (Tacon and Metian, 2013). Notwithstanding Germany, pelagic fish assume a significant part in consumption for all non-industrial nations for the situation considers. The accessibility of marine fish in what's to come is vital, particularly for those nations with low aquaculture production.

Key Takeaways

- Total fish consumption has expanded after some time, however, the portion of marine fish also, fish have declined.
- Marine profits still assume a significant part in fish consumption, a few nations depend on wild profits at 100%.
- Increasing fish consumption is driven mostly by Chinese aquaculture.
- Aquaculture items are not a replacement choice for all nations.

Worldwide Importance of Fish as a Protein Source A remarkable blend of top-notch protein and indispensable supplements make fish a significant food. Fish isn't just a source of animal protein – 150 g of fish gives around 50 to 60% of a grown-up's every day protein necessities – yet in addition of unsaturated fats, nutrients and other fundamental components like iodine and selenium, which don't happen in such amount and variety in cereals, different harvests or meat.

Dietary variety and dietary ampleness are high on the plan in the battle against craving and unhealthiness. Extreme neediness is exceptionally associated with the requirement of 'too much' staples and too little protein, fat and micronutrients. In coast locales of non-industrial nations, fish is regularly the solitary moderate and generally effectively accessible source of animal protein. In nations like Sierra Leone, which has an extremely low food security, the portion of fish in animal protein is more than half. Checked contrasts exist between and inside nations and districts as far as the amount consumed per capita and the resulting commitment to healthy requirement. In a worldwide correlation, Africa and Latin America burn-through moderately little fish (around 10 kg for every individual each year), though per capita consumption in Asia, North America and Europe are over the worldwide normal (20 kg) at around 22 kg each year (Shah et al., 2018).

This mirrors the components that influence fish consumption: how accessible fish is, how costly it is, regardless of whether there are dietary practices comparable to fish and how fostered the nation is. As a rule, the lower the pay the lower is the fish consumption. The dietary practice identifies with the way that nations with a solid fishing custom because of a long coastline, many fish-rich streams or islands will, in general, consume more fish. The WHO suggests on normal a yearly requirement of 11.7 kg fish per individual (around 32 g each day or 225 g each week). Found the middle value of overworld locales, just Africa and Latin America verge on gathering this reference value. Universally talking, be that as it may, there is an inconsistent dispersion of fish and the Northern side of the equator burns through an excess of fish for each capita. In 2013, fish represented 6.7% of all protein burned and for 17% of the worldwide population's requirement of animal protein. In non-industrial nations, this offer was 19.6% and 24.7% in low-pay food-shortage nations. For 3.1 billion individuals, fish represents 20% of their animal protein; for 4.3 billion individuals, this offer is 15%.

Some little island states, for example, Kiribati, Micronesia and the Maldives depend solely on fish as a protein source. The everyday dietary commitment of fish as far as calories are around 34 calories for each capita. In nations where elective protein food is absent and where there is a traditional inclination for fish (for example Senegal), just as in a few little island states, for example, the ones named over, the everyday fish calorie requirement arrives at 130 calories for each capita or more. Be that as it may, in light of this information, we may think little of the significance of fish as a protein and supplement supplier, specifically for nations with low food security or potentially poor populations. There are various explanations for this.

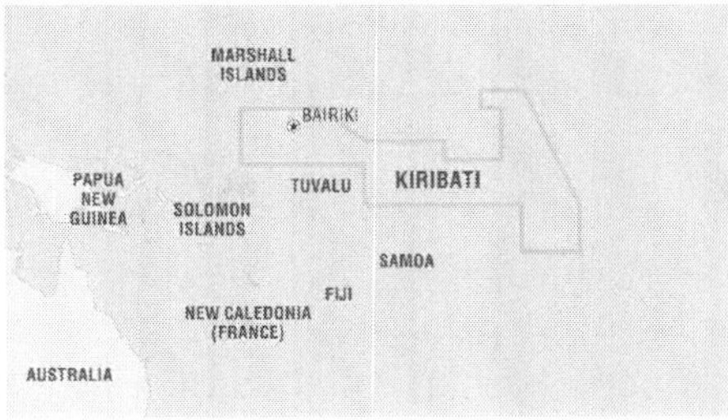

Figure 1.6: Kiribati.

Source: https://www.britannica.com/place/Kiribati

- There is a tremendous variety between and inside nations, specifically seaside areas inside little island nations depending to a lot further degree on fish as a source of animal protein. At the point when these coast areas are far off, for example far away from significant business sectors and not effectively open, and have a high predominance of destitution, replacement prospects are restricted in the short run. Consumption for this situation is supply-driven.
- Consumption information is probably going to be disparaged considering the under-recorded commitment of resource fisheries and small-scale fisheries in true measurements. Henceforth, real fish consumption in non-industrial nations is presumably higher than true information reports.
- Economic reliance on fish as a kind of revenue assumes a significant part in marine regions in agricultural nations with repercussions on food security.

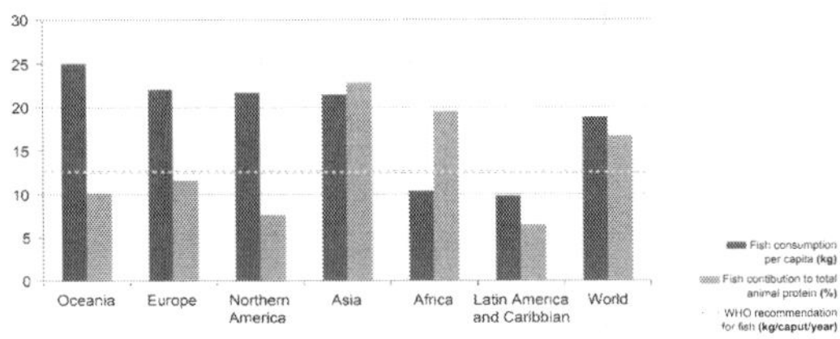

Figure 1.7: Total fish consumption and fish contribution to total animal protein.

Source: https://www.fishforward.eu/wp-content/uploads/2017/11/future_trends_report-EN.pdf

The dietary example of needy individuals ordinarily has an exceptionally solid segment of staple food (specifically maize, rice and different oats), with fish consumption a significant factor in assisting with rectifying an imbalanced calorie/protein proportion. Fish frequently addresses a moderate source of animal protein that may not exclusively be less expensive than other animal protein sources yet in addition part of the neighborhood or potentially traditional plans. In nations with a long coastline, for example, Senegal and islands like Indonesia, fish represents or surpasses, 40% of absolute animal protein consumption (Wurts, 2000).

The figure underneath we give a graphical portrayal of the job of fish in the eating regimen according to financial turn of events. While extremely low pay will in general be related to hunger, rising pay first prompts sufficient accessibility of calories and hence to satisfactory nourishing quality. At low pay levels (base left) fish adds to a irregular eating routine: fish will, in general, be burned- in minuscule or extremely enormous amounts relying upon accessibility. At exceptionally major league salary levels (upper right), fish additionally will in general add to an unequal, excessively protein-rich eating regimen. Monetary improvement is related to the underlying change in which the portion of individuals associated with fishing decreases following expanding advancement.

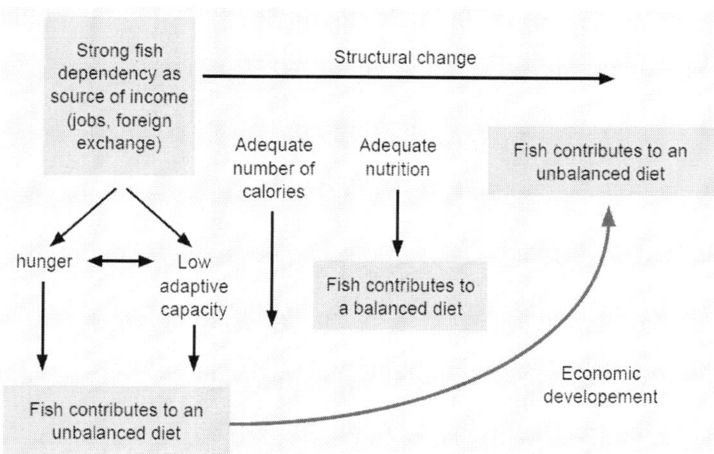

Figure1.8: Importance of fish in the diet concerning the economic development of a country or population.

Source: https://www.fishforward.eu/wp-content/uploads/2017/11/future_trends_report-EN.pdf

While in outright numbers animal protein requirement is lower in non-industrial nations than in developed nations, the developing portion of animal protein overall is driven principally by agricultural nations getting up to speed, especially in Africa and Asia. Separating animal protein requirement into fish and meat, the commitment of fish to add up to animal protein consumption has been declining marginally since 1990 over other animal proteins.

In the global south, nations, where fish is a conventional food, expanding pay, prompts expanded meat consumption and more noteworthy consumption of more important fish species, for example, demersal rather than pelagic fish. Of course, the consumption of fish from aquaculture is developing quickly. Somewhere in the range of 1990 and 2012, fish consumption from wild sources remained practically something very similar, however, consumption of fish from aquaculture has duplicated by five. A big part of fish delivered for human consumption came from aquaculture in 2012, contrasted with only 5% in 1962 and 37% in 2002. Similarly, as there is sufficient food on normal to take care of everybody in the world and appetite is an issue of circulation, there is additionally no protein hole considering current world normal protein supply.

Around the world, the normal protein supply in 2011 was 79 g for every capita each day, while the normal protein need was 49.6 g per capita each day. The last was determined from the suggested 0.8 g per kg of body weight and the normal load of an individual (62 kg) in 2011.

1.4 FOOD SECURITY AND FISH

Undernourishment is a significant issue around the world, with one out of seven individuals undernourished and more than 33% of baby mortality owing to undernutrition. This is particularly the situation in many agricultural nations, with the majority of undernourished individuals living in country regions. A large portion of the world's undernourished individuals live in South Asia, firmly followed by Sub-Saharan Africa and East Asia. Notwithstanding the predominance of undernourishment, specialists frequently utilize a few different markers to survey the food security level of a country like hindering, estimated by stature for age, and wastage, estimated by weight-for-age.

Figure 1.9: East Africa fish farming project supports food security, mitigates climate change.

Source: https://www.seafoodsource.com/news/aquaculture/east-africa-fish-farming-project-supports-food-security-mitigates-climate-change

A typical multidimensional file utilized is the Global Hunger Index (Shown in the table underneath). To mirror the multidimensional idea of appetite, the GHI joins the accompanying four-part pointers into one file:

- 1/3 for undernourishment: the extent of undernourished individuals as a level of the population, mirroring the portion of the population lacking caloric requirement.
- 1/6 for kid wastage: the extent of youngsters younger than five who experience the ill effects of wastage, for example, low weight for their tallness, reflecting intense undernutrition.
- 1/6 for youngster hindering: the extent of kids younger than five who experience the ill effects of hindering, for example, low stature for their age, reflecting constant undernutrition.
- 1/3 for kid mortality: the death rate of kids younger than five, part of the way mirroring the lethal collaboration of insufficient sustenance and unfortunate conditions (all normalized dependent on edges set marginally over the most elevated nation-level qualities noticed worldwide for that pointer somewhere in the range of 1988 and 2013.

This pointer underscores the nourishment circumstance of kids – a weak subset of the population for whom an absence of dietary energy, protein or micronutrients (for example fundamental nutrients and minerals) prompts a high danger of disease, poor physical and intellectual turn of events or death. It additionally joins autonomously estimated markers to decrease the impacts of irregular estimation mistakes. In our respective nations, Senegal and Indonesia fall into the real class, South Africa has moderate levels, and all others are in the low classification. The record isn't determined for Germany, France and the USA, which are viewed as for the most part food secure.

	1995	2005	**2015**
	With data from		
	1993-1997	2003-2007	**2010-2016**
China	23.2	13.2	8.6
Indonesia	32.5	26.5	22.1
Senegal	36.9	28.5	23.2
South Africa	16.5	21.0	12.4
Peru	25.0	18.8	9.1

Figure 1.10: Global Hunger Index 1995 to 2015 for case study countries.

Source: Welthungerhilfe (WHH); International Food Policy Research Institute (IFPRI); and Concern Worldwide 2015.https://www.fishforward.eu/wp-content/uploads/2017/11/future_trends_report-EN.pdf

1.5 RELIANCE ON FISH

In this part, we consolidate the country-explicit circumstance of food (in) security and general wellbeing/hunger circumstances with the financial value of fish and fisheries to the occupations in these nations to portray the nations' reliance on fish. The accompanying components decide reliance on fish for food and nourishment:

- Fish as a food source, for example as a source of calories. This is especially important in nations like Senegal with food limitation and high fish consumption.
- Fish as a protein and micronutrient source. This is especially applicable in nations where the protein portion of fish in the eating regimen is high or low. It is likewise especially significant in practically all poor countries that have an imbalanced eating regimen, where individuals eat an excess of staple food and insufficient micronutrients (for example Senegal, Indonesia). The fisheries and aquaculture area are vital in extended and sound

eating regimens. While normal per capita fish consumption might be low, even little amounts of fish can have a critical positive healthful effect, given that it is a concentrated source of fundamental dietary parts. Micronutrient defiencies influence countless individuals, especially ladies and youngsters in the growing stage. Over 250 million kids overall are in danger of nutrient A deficiency (prompting visual impairment), over 30% of the total population are iron deficiency, 200 million individuals have goiter with 20 million experiencing learning challenges because of iodine insufficiency and 800,000 youngster mortalities each year are owing to zinc inadequacy (Yang et al., 2020).

Economic reliance on fish as a pay source to purchase (good) food. This is especially important in nations with high destitution rates and in nations where individuals working in fisheries are similarly poor. At the point when they lose their positions in the fish business, destitution rises and this influence slims down: craving and nature of the eating regimen. An absence of cash to purchase food drives first to a less adjusted eating regimen, with a propensity to eat less fish and meat, subsequently less protein and micronutrients and in much more extreme cases to a lower requirement of calories generally. The seriousness of this impact relies upon the strength of the formal or casual government-managed retirement systems set up. For instance, if Senegal quit fishing totally, this would unfavorably affect the vocations of fishermen with a far and wide appetite as a probable result, essentially in the short run. This is an ordinary appropriation issue: around the world talking, enough is accessible, however, it is circulated profoundly inconsistent.

Thus, fish adds to the dietary security of helpless families in non-industrial nations differently. These incorporate a consumption pathway where the immediate consumption of fish expands requirements of calories, yet more critically protein, micronutrients and omega-3 oils, and a money pay pathway, whereby the fish business adds to work and higher by and large food consumption in poor nations. Commercialization, fish handling and small-scale aquaculture offer significant work openings, especially for ladies in non-industrial nations, through their immediate association in the production, preparing an offer of fish. Specifically, nations, where an enormous portion of the population is subject to fishing (for example Senegal), may experience the ill effects of higher appetite rates and political insecurity if the fishing area decreases. As per estimates by the FAO, an aggregate of 660 to 820 million individuals are straightforwardly or in a roundabout way reliant upon fisheries and aquaculture. The FAO appraises the number of fishermen alone at 54 million, of which 87% live in Asia. In non-industrial nations, most of them work in small-scale fisheries with low fish production per individual: on normal 1.5 tons, each year contrasted with 25 tons for every angler and year in Europe.

Figure 1.11: Main drivers of reliance on fish.

Source: https://www.fishforward.eu/wp-content/transfers/2017/11/future_trends_report-EN.pdf

1.6 FISH DEPENDENCE

Research estimates the level of reliance on fish as a kind of revenue and food, specifically as a source of protein. It depends on a composite arrangement of the accompanying aspects:

- food security: the pervasiveness of undernourishment for the period from 2011 to 2013 in %, information from the FAO;
- fish consumption: fish portion of all-out animal protein requirement in % for 2011, information from the FAO;
- catch production per capita for 2011;
- (GDP) per capita as an intermediary for replacement limit with regards to 2011, information from the World Bank.

The pointer estimates the short-run reliance. Over the long haul, the opportunities for repaying through new businesses (for example developing a aquaculture industry) and new sources of protein (for example plant-based sources) must be thought of. The principle distinction is the later information utilized in this file and the more grounded centre around fish as food.

Utilizing later information (from 2011) is just conceivable because this file is made out of fewer parts. The aftereffects of the two records are as a rule the same. Fish as a portion of animal protein in food and catch production as an intermediary for monetary importance are the primary factors and weighted similarly. Gross domestic product per capita (in USD) and food security (predominance of undernourishment) are the directing elements. A major league salary decreases the reliance made by an undeniable level of the two principal factors because of remuneration impacts. This implies that a big-league salary builds opportunities for making up for a deficiency of occupations or the deficiency of fish as a source of food, for example through imports and federal retirement aid. A significant degree of food security diminishes the possibly hurtful impacts of losing fish as a food source (Yang et al., 2020).

Information and Estimation of Model Parameters We use information from three fundamental hotspots for the assessing boundaries of the bio-financial fishery models:

- Sea Around Us (www.seaaroundus.org),
- FAO information base FAOStat (http://faostat3.fao.org/home/E) and
- FAO Fisheries and Aquaculture Department's information base FishStatJ (http://www.fao.org/fishery/measurements/programming/fishstatj/en).

The Sea Around Us data set remembers separated information forgets, arrivals and assessed disposes of enormous scope, small-scale, resource and sporting fishery. The information in the Sea Around Us data set joins revealed values from the FAO and extra unreported information assessed via Sea Around Us. Extra data on missing information was gathered for this gauge. The fundamental sources were administrative sites and distributions, factual offices liable for the fishing business, worldwide examination associations like the FAO, the International Council for the Exploration of the Seas (ICES) or local fisheries the board associations like the Northwest Atlantic Fisheries Organization (NAFO) just as scholastic writing. Given this data, anchor-focuses in time are determined and utilized for the introduction. A direct insertion between anchor-focuses is utilized to recreate business profits. Either population patterns or patterns in the number of fishers over the long haul are utilized to insert between anchor focuses for non-business profits from resource and sporting fisheries. The recreated profits are then joined with the detailed FAO information. The Sea Around Us get information is disaggregated into profits by species, utilitarian gatherings and size gatherings of 0-30, 30-89 and >89 cm normal length. We utilize this data to bunch the profits into predator and prey classifications. We adopt a size-based strategy and believe huge fish to be doubtlessly predator and little fish to probably search fish. In particular, we consider all fish more noteworthy than 90 cm to be predator while fish more modest than 90 cm and spineless animals are viewed as prey. In the worldwide predator prey model and the worldwide interest model, the information is collected each year over every one of the 64 large marine ecosystems (LMEs), prompting 61 observations each (= number of years from 1950 to 2010) for predator and prey profits. In the fish supply model at the LME-level we utilize all-profitseach year collected oversize gatherings. In this model, the information has been totaled by LME and year. Altogether, we utilize a dataset with 3,904 observations. The figure below shows the improvement in absolute profits and profits from 1950 to 2010. The profitsin 1996 is at 123 million tons. From that point forward, profits are diminishing. This is per scholars who likewise determined the worldwide top in profits in 1996 at 129 million tons.

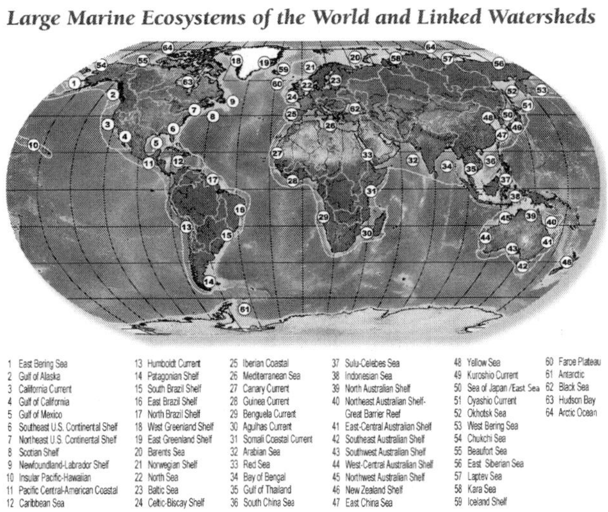

Figure 1.12: Large Marine Ecosystems.

Source: https://www.researchgate.net/publication/26798346_Accelerated_Warming_and_Emergent_Trends_in_Fisheries_Biomass_Yields_of_the_World%27s_Large_Marine_Ecosystems/figures?lo=1

Rather than the worldwide information of these experts do exclude high oceans catches, which brings about the distinction of 6 million tons. Furthermore, some little island states, like Wallis and Futuna Islands (France), Saint Helena (UK) or Nauru, are excluded because their geological position doesn't exist in the space of an LME. The natural boundaries are assessed utilizing the Catch-MSY technique created by Martell and Froese. This strategy permits natural boundaries, for example, biomass dependent on catch information, to be assessed. It requires time series of catch information and earlier ranges for the boundary values as well as could be expected scopes of stock sizes from the underlying and the last time frame. In the wake of indicating starting stock sizes and cutoff points for the last stock size, a boundary set is arbitrarily drawn from the earlier boundary dissemination. Then, at that point, the fundamental fish supply model is utilized to ascertain the biomass comparing to the degree of the collection given the boundary set. If this biomass is in a sensible reach, the boundary set is put away. In our investigation, we rehash this technique multiple times for each LME. We use tests of 1,000 haphazardly picked acknowledged boundary values in our model calculations to process mean evaluations and certainty stretches. Accordingly, all outcomes revealed below depend on midpoints and standard deviations acquired from 1,000 separate model runs (Zhang et al., 2015).

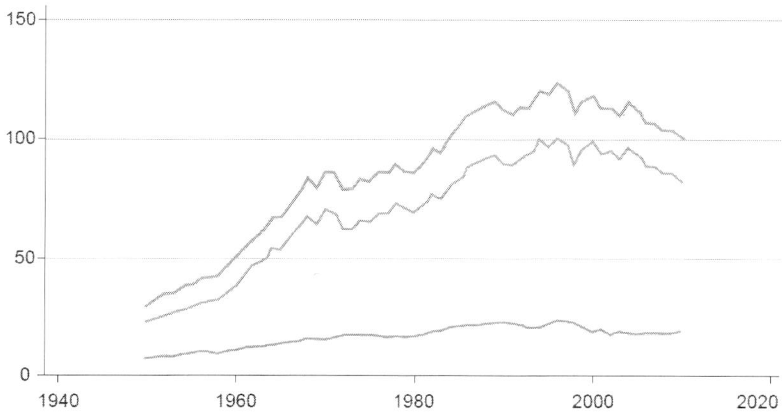

Figure 1.13: Development of worldwide total profits and predator/prey profits from 1950 to 2010. (in million tons) Source: SeaAroundUs.

Source: https://www.fishforward.eu/wp-content/transfers/2017/11/future_trends_report-EN.pdf

We utilize the previously mentioned approach by Martell and Froese with a Schaefer overflow production model for each LME and one worldwide supply of collected fish. In the worldwide predator prey model, we expand the methodology by Martell and Froese and decide boundary values for a Lotka-Volterra predator prey model. For each situation, introductory boundary sets are arbitrarily drawn from a uniform dispersion to be tried. The financial hypothesis predicts a positive connection between fish stock biomass and market supply of fish (or no relationship at all on account of an unadulterated tutoring fishery), and consequently a negative connection between stock biomass and the fish cost. We use value information in each run for a tried boundary set to check whether this prerequisite is met. Natural boundaries that don't finish this assessment are dismissed. Else, we utilize the subsequent data on the connection among cost and stock biomass to acquire an approximate for financial boundary values.

1.7 GLOBAL AND REGIONAL DEMAND SYSTEMS

To evaluate the interest systems, we use the information on fish costs from Sea Around Us. The information on landed qualities is utilized to infer costs. The ex-vessel fish costs, used to ascertain the landed qualities, are inferred utilizing two methodologies. Nearby ex-vessel costs, changed over to US dollars, structure the beginning stage. These are joined with ex-vessel costs determined from detailed landed qualities and profits. Genuine costs are controlled by emptying utilizing the purchaser value record of 2005. We utilize this valuable information for the computation of production values in the worldwide interest model. Since prey likewise incorporates significant spineless animals like shrimp, lobster or ocean imps, the prey cost doesn't go astray much from the predator cost. Notwithstanding fish production amounts and fish costs, the worldwide interest model likewise requires information on complete uses and utilization levels for the three products: ruthless fish, prey fish and non-fish protein-rich replacement merchandise. Complete public uses are determined from production, price and import value, while public utilization is determined from production, price and import amounts.

Information on replacement merchandise is taken from the FAO Statistics division, with a period series running from 1961 to 2013. Assortment of the information is confined by the accessibility of both exchange and production information and the length of the relating time series. Thinking about this load of restrictions, the accompanying items are remembered for the gathering of replacement merchandise: beans, dairy items, eggs, lentils, peas, maize, meat, nuts and rice. The worldwide interest model reflects absolute worldwide production (amount in tons), worldwide price cost (per ton in current USD) and worldwide price value (in current USD) for the previously mentioned products. As opposed to this, the interest model at LME-level doesn't need information on predator fish and prey fish, yet on locally created fish and imported fish. We utilize the FAO data set to give information on absolute public homegrown production (amount in tons), sends out (amount in tons, value in current USD) and imports (amount in tons, value in current USD) from 1976 to 2010 (Sandersen & Kvalvik, 2015).

The FAO FishstatJ data set doesn't contain data on production values for fish merchandise subsequently we figure the production value as the result of production amounts and price costs. Price (import) costs for fish just as for replacement merchandise are determined by partitioning trade (import) values by the relating amounts. Ocean Around Us just gives emptied landed qualities (genuine costs) of ruthless and search fish and is consequently not tantamount to the declared FAO value information for nonfish replacement merchandise. Therefore, information on worldwide price values and worldwide price amounts of fish are additionally taken from the FAO FishStatJ data set for the period from 1976 to 2010. Worldwide ostensible price costs per ton are determined utilizing this information. The FAO FishstatJ data set doesn't separate scavange and predator fish. Therefore, we couldn't compute the cost for these two kinds of fish independently. All things considered, a typical worldwide price cost each year for the two kinds of fish is determined. Besides, both wild-caughtfish and fish from aquaculture are remembered for the FishstatJ data set.

Since we ascertain a worldwide cost for all fish wares in this information base, this cost doesn't contrast between wild-caughtfish and fish from aquaculture all things considered. Henceforth, the data went into our data set from these two sources identifies with:

- complete worldwide marine profits of predator and scrounge fish (in tons),
- the worldwide price value of predator and prey fish (in USD),
- the worldwide price cost for fish (per ton in current USD).

At long last, absolute public uses are determined as the amount of production and import values less price value. Public utilization is determined as the amount of homegrown production and imports fewer prices. For certain observations, the subsequent incentive for utilization is negative. In these cases, we set the negative qualities equivalent to nothing. We expect one worldwide customer who has inclinations concerning the amounts of every item she burnsthrough. We further expect that the buyer likes to substitute fish with another type of fish instead of with non-fish substitute merchandise. To figure worldwide interest for a) predator fish, b) scavenger fish and c) protein-rich non-fish materials, we utilize the accompanying yearly information at the worldwide level: trade costs, production amounts, absolute uses for each of the three products and a boundary that communicates the previously mentioned inclinations of our purchaser.

With the given data on predator and search fish and substitute merchandise, we gauge the interest boundaries for every product for every year from 1976 to 2010. The interest boundaries demonstrate the assessed share every ware has in the utilization of protein-rich food. For the projections of worldwide fish interest in 2050, we then, at that point compute the mean of these interest boundaries after some time per LME and item. To coordinate with getting information from Sea Around Us at the LME level with FAO country information, we assessed every nation's offer in a specific LME dependent on the spatial cross-over of the country's seaside waters with the LME.

A few nations remembered for the FAO information don't have marine waters in any LME, either because they are landlocked or because there is no LME characterized in their coast waters. We eliminate nations without access from our dataset.15 Some nations have coast waters in more than one LME. For these nations, we expected that the small amount of exchange and production related to an LME is equivalent to the negligible part of the country's srate of coast waters in the comparing LME. The offers are determined utilizing GIS programming and data on the srate of selective financial zones (EEZs) and LMEs. Assessing local utilization requires the conglomeration of all info information (homegrown production amounts, and so forth) at the LME level. Concerning the locales, we bunch input information by the 64 LMEs given via Sea Around Us. In each LME, the shopper has inclinations in regards to the amounts of every ware she burns through.

We accept that each LME's purchaser likes to substitute fish with another type of fish as opposed to with a non-fish substitute great and that she separates among imported and locally delivered fish. To ascertain local interest for a) imported fish, b) locally delivered fish and c) protein-rich non-fish substitute merchandise, we utilize the accompanying yearly provincial level info information: trade costs, import costs, homegrown production amounts, imported production amounts, absolute consumptions for every one of the three products and two boundaries that express the previously mentioned inclinations per LME. Utilizing this data, we gauge the interest boundaries for every great and every year from 1976 to 2010 (Rosas et al., 2019). The interest boundaries indicate the assessed portion of every one of the three products in the absolute utilization of protein-rich food varieties. To extend the interest for fish in 2050, we figure the normal value per LME and per great from the time series of the interest boundaries.

Chapter 2

Major Species Reared in Extensive Aquaculture

CONTENTS

2.1 INTRODUCTION

Due to the increase in demand, aquaculture has become the fastest-growing food-producing industry globally. Aquaculture involves cultivating aquatic organisms mainly for human consumption. It is an alternative source of seafood due to the high levels of over-fishing in the natural resources. Fish farming has proven to be a sustainable option for consumers as fish has the highest protein retention compared to other sources making it highly resource-efficient. On top of that, it has the lowest feed conversion ratio. It has the lowest greenhouse gas emissions compared to other forms of farming.

Fish farming will help improve the health of the planet and that of the population. It is expected that the global population will hit 50 billion in 2050. This will increase the demand for protein but at least 52%. This being said, fish as a source of food is of high economic importance to the fishermen and even to the global economy.

Fisheries provide jobs to over 820 million people globally. There is a large variety of jobs, from the production and sale of inputs (vessels, fishing gear, bait, etc.) to farming and harvesting, processing, marketing and distribution of fish. Most of the working community is isolated in rural areas with a lack of infrastructure and even knowledge and information. Due to the lack of knowledge on the management of fisheries, they suffer high post-harvest losses of already low volumes of production. These are caused by improper handling and processing coupled with a lack of proper storage facilities and distribution technology (Naylor et al., 2021).

Extensive pond farming is a traditional extensive freshwater fish farming that is practiced in the whole of Europe, especially in Central and Eastern Europe. It involves maintaining ponds, that can be natural or artificial, so as to ensure the development of aquatic fauna. During the winter, the ponds are thoroughly cleaned and fertilized so as to promote aquatic vegetation and in turn increase the availability of micro-organisms, small mollusks and crustaceans and larvae and worms. These form the basis of the aquatic food pyramid. This helps to promote the culture of profitable fish at a yield that is higher than that of the ecosystem.

In contrast, the intensive freshwater fish farming site is mainly made up of a couple of open-air concrete tanks, raceways or earth ponds each with different sizes and depths. They are made like this to suit the different stages of fish growth. It is used typically for trouts.

In Marine fish farming, they use extensive brackish water farming and intensive sea farming. Depending on their geographical situation, the lagoons and coastal ponds provide sea-bass, eels, sea-bream, mullets, sturgeons, crayfishes and shellfish. In intensive sea farming, they use sea cages that hold fish captive. The fish are held in large pocket-shaped nets that are supported at the bottom and kept on the surface by a framework that floats and is either rectangular or circular. It involves several cages that are in groups together in rafts. They have walkways that are for boat access, storage of feed and equipments for feeding the fish. It has an open system that makes it susceptible to external influences such as pollution and physical impact.

European countries also practice Shellfish farming. This is based primarily on species that are born in the wild and get their nutrition from the environment. No input is required because they feed on planktons which they filter through their gills. There are a couple of techniques that are used. These include bottom farming, which is often practiced in shallow coastal areas of up to 10m deep, inter-tidal shellfish farming which uses areas between high and low tide, and floating systems such as rafts and longlines, which can be used in open sea or estuarine environments (Milstein, 2005).

Its potential brand value, growth rates and the ability of a species to operate under circumstances of culture are the main criteria that define the acceptability of a species. Even more crucial for an aqua culturalist than biological elements is economic concerns for the selection. Carnivorous types usually require a high diet of protein and are consequently regarded as more expensive to rear, however, expenses are mostly dependent on the local availability and prices of the necessary feed. However, the more carnivorous species have higher market prices in compensation for high feeding expenditures.

In general, such species have increased market potential, thereby attracting significant investments. For most aquaculture operations, this is particularly important for catch-based aquaculture, as most of the species cultivated are carnivorous. The species groups of mollusks (scallops, oysters, mussel or scallops), crustaceans (shrimp, crabs) and fin fish are utilized in catch-based aquaculture (eels, grey mullets, milkfish, yellowtails, groupers, rabbit fish, tunas). Four target categories of species are included in this report: yellowtails, tunas, groupers, and eels. The latter have been chosen for their importance in captured aquaculture because they are rapidly growing and have significant market demands; their culture and technology advancements are also increasingly interested to support their upbringing.

Below are fish bred through aquaculture:

2.2 FRESH WATER FISH

a. Indian Major Carps

- *Catla catla*: This fish has a deep body and a very prominent head. It can grow to a size of 1.8 meters and weight of 45kgs. It uses large gill rakers to feed on zooplanktons on the surface of the pond. In addition, the younger ones also feed on phytoplankton. It is one of the fastest-growing fish. In the first year it grows to about 35 to 45 cm and has a weight of 1.5 to 2 kgs. The second-year is its year of maturation.
- *Labeo rohita*: This carp is considered the tastiest carp of them all. It has a small pointy head with a small mouth and a fringed lower lip. It is distinguished by its full reddish scales. It feeds on phytoplankton, decaying debris and plant debris. The younger ones only feed on zooplankton. It grows to a maximum size of 1 meter. It grows fairly fast with the first year growing to 35 to 40 cm and weight of 900 grams.
- *Cirrhina mrigala*: This fish has a linear body with a small head. It has thin non-fringed lips. Its body is bright and silvery with a golden tinge. It is a bottom feeder and feeds on decaying organic debris or vegetable debris. It grows to a maximum size of 0.9 meters. In the first year, it can grow to 30 cm and weight of 700 grams.
- *Labeo calbasu*: This carp is quite popular and widely distributed. It grows to a length of 30 to 35 cm and weight of 450 grams in the first year.

b. Catfishes (Order: Siluriformes-'Keluthi')

Catfish are capable of breathing the air from the atmosphere making them air-breathing fish. This ability allows them to be transported live, even for long distances, and sold in a fresh condition to the market because they can survive for a long time without water. They have an upper and lower jaw that have two pairs of long barbels each. They have no scales. Most catfish are considered predators and cannibals. They feed on all pond animals (Asche et al., 2013).

They are farmed in warm regions and are available to the market at an affordable price. They have stings embedded in their fins therefore one needs to be careful when handling them. Their flesh is not as white as you'd expected compared to other fish but when cooked well, they can be very tasty.

c. Murrels or Snakeheads (order: Channiformes-'Viral' meen)

These fish have a high demand. They are also considered air-breathing fish. They have cylindrical bodies that are elongated with a depressed head and large protractile mouth. They can be bred and cultured in shallow swamps and irrigation wells. *Channa* spp are one of the common staple foods in Vietnam, Thailand, Cambodia and South east Asian countries. They are mainly carnivores. In addition to being a staple food, they are also used for their therapeutic nature. They are used for healing wounds and reducing post-operative pain. They are also sold in the ornamental fish industry.

- *Channa marulius* (Giant snakehead): This species has no spines on the fins. It can grow to a length of 1.2 meters. It can be cultured in ponds together with tilapia. Young tilapia are a source of food to the Giant Snakehead.
- *Channa striatus* (Common Murrel): It is distinguished by stripes on its body. It grows to a length of 90 cm maximum.

d. *Tilapia* (order: Perciformes)

Oreochromis mossambicus: This is an exotic fish that originates from the east coast of Africa. It has a soft dorsal fin posteriorly and dorsal fins with spines anteriorly. These fish can mature even at two months and they breed almost eight times in one year. The female guards the eggs that are fertilized in her mouth.

e. Trout.

These are normally found in ponds, lakes or cool clear streams. They can also be cultured in fish farms then transferred after maturity into freshwater lakes. They feed on other fish. They are bony and have very tasty flesh due to their feeding habits on crustaceans.

f. Bluegill.

This fish originates from North America. It can be found in lakes, ponds, rivers and streams. It can survive in both shallow or deep waters but tends to live next to underwater structures or even the stamps of trees. They like to find shelter under aquatic plants or even any shade from trees along the bank of the river. They grow to a maximum length of 16 inches. Their flesh is white and flaky and has a sweet taste (Allen et al., 2012).

g. Large mouth bass.

These fish are much sought after, not only for food but also for the sport of fishing. Fishermen like the challenge because they have resilience and it causes excitement. They go airborne when caught on a hook to try and unhook themselves. They are found in shallow water covers and around aquatic plants. The larger fish breed the females, so it is advisable to catch the smaller fish. Bass that are cultured in clear, cold and clean water taste the best.

h. Salmon.

This is a very common fish and is farmed all over the world. Salmon is rich in Omega-3 acids, Vitamin D and proteins. Due to their oily nature, they are rich in Cholesterol as well. They have red or orange flesh but it is also possible to find white flesh. They are carnivores. In the Atlantic, most salmons are farmed while in the Pacific, most are caught from the wild.

i. Walleye.

This is one of the best-tasting freshwater fish. They are found at the bottom of lakes and streams. These are a bit harder to catch. They come closer to the surface during the early morning hours or the late evening hours.

j. Cod.

These are found both in pacific and Atlantic waters. In the Atlantic, they are found deep in the cold waters of the North Atlantic. They are found in the eastern or western parts in the pacific. Their flesh is white and flaky. It is a staple in places like Brazil, Italy and Spain. The livers of this species are processed to produce Cod Liver oil. Cod liver oil is rich in many nutrients including Vitamin A, D and E and Omega-3.

2.3 MARINE FISH

a. Nethili/Indian Anchovies.

These are small oceanodromous fish from the family Engraulidae.

They are small and green with silver stripes that are longitudinal. They can grow to a length within the range of 2 to 40 centimeters. They eat plankton any recently hatched fish. They are found in temperate waters. They do not like very cold or very warm waters. They are often used as dead or live bait to catch tuna.

b. Rawas/Indian Salmon.

It is originally named *Eleutheronema tetradactylum*. It is found in the shallow muddy areas in the ocean. The adult fish feed on prawns and fish. Sometimes they feed on polychaetes. The younger fish feed on shrimps and prawns. Occasionally they feed on mysids. It is a favorite in India as it is very tasty (Mazur & Curtis, 2008).

c. Bhetki/Barramundi.

It is also called the Asian sea bass. It is a catadromous fish which is found in the Pacific region. It has a long body with the upper jaw extending behind the eye. It has large silver scales which change in different environments to become darker or brighter. They can grow to a length of 1.8 meters and a maximum weight of 60 kgs. They like temperatures of 26-30 degrees Celsius. They have white flaky flesh with a considerable amount of body fat. It is one of the most expensive fish.

d. Butterfish/Pomfret.

These are perciform fish in the family Bramidae family. They are found in the Pacific, Indian and Atlantic Oceans. They have a characteristic flat body with forked tail fins. They come in either white or silver color. They can grow to a weight of 4 to 6 kgs. It is known to have a very good taste with buttery soft flesh.

e. Surmai/Seer-fish.

This is a fish of the sea: of the Mackerel variety. It is found in the Indian ocean. It can grow up to a weight of 45kgs. It is cooked and eaten but also has another purpose. It is used to make a delicacy called fish pickle, which is eaten with rice as a condiment. It is also a game fish and is known for jumping out of the water when hooked to unhook itself.

f. Bangda/Indian Mackerel.

This fish is found in the Indian and west pacific oceans. It likes to be in the warm shallow waters. Its head is longer than its body depth and has thin dark golden bands on the upper part of its body that run longitudinally. It has a black spot on the lower margin of its pectoral fins. The Bangda fish grows to a maximum length of 35 cm and prefers temperatures of 17 degrees Celsius.

g. Tarali/ Indian oil Sardine.

Tarali is a ray-finned ship that is found in the genus Sardinella. Together with the mackerel, they form the two most important fish in India. It feeds on diatoms and copepods. It has an elongated body with a round belly and 8 rays on its pelvic fins. Its mid-lateral line is golden making this a distinct character.

h. Bangus/Milkfish.

This fish can grow to 1.8 meters long but many of them only grow to 1 meter long. It can reach a maximum weight of 14 kgs. Its body is elongated and seems compressed with a small toothless mouth. It has an olive green color with bordered fins that are darkly colored. It feeds on algae and small invertebrates. It is an important source of food in Indonesia, India and the Philippines.

i. Mangrove red snapper.

This fish is from the snapper variety. It is originally found in the Indian Ocean and the Pacific ocean. It can be found in many different colors: copper, bronze, burnt orange and reddish-brown. The different colors are depending on the age of the fish and the environment it lives in. Younger fish have been noted to be darker. It has large canine teeth that they use to hunt for its prey. It is a carnivorous fish and feed on crustaceans, other fish and cephalopod mollusks. It is highly regarded for its sweet white flesh that is firm.

j. Bombay Duck.

It is a lizard fish also called Bummalo. It can grow to a length of 40 cm but is commonly found at 25 cm. It resides in the Indo-Pacific tropical areas. Its flesh has no specific taste and is therefore dried and salted before it is eaten. Drying it causes it to have a very strong odor and because of this, it has to be transported in an airtight container.

k. Tuna fish.

This saltwater fish is from the Thunnini tribe. It is one of the only fish together with Opah and Mackerel Sharks that can maintain a body temperature that is above that of the surrounding water. It is an agile predator and among the fastest swimming pelagic fish eg. The yellowfin tuna can swim up to a speed of 75 km/hr. It prefers to reside in warm waters. It is well streamlined with a metallic blue dorsal side and silver or whitish ventral side. These colors help it camouflage (Maulu et al., 2019).

l. Pangas Catfish.

This comes from the shark catfish variety. It grows to a length of 3 meters. It is found in the brackish waters in India and also in fresh and seawater.

m. Herring.

These are forage fish that moves in large schools near the coast. Culpea is the most common species is found in the North Pacific and North Atlantic oceans in their temperate shallow waters. They are silver in color and have a single soft dorsal fin without spines. They have lower jaws that notably protrude. In addition to being harvested for food, they are also harvested as baitfish. They have been a staple food since 3000 BC. They have high amounts of Omega-3 acids making them very nutritious.

n. Mahi-mahi.

Also called the dolphin fish. They ate known for their dazzling colors: golden on the sides, and bright blues and greens on the sides and back. Females are normally smaller than the males weighing around 7-13 kgs. They are carnivores and feed on crabs, squid, mackerels, etc. and are one of the fastest-growing marine fish.

Below we will discuss the 10 most commonly bred fish in intensive aquaculture:

2.4 BARBELS, CARPS AND OTHER CYPRINIDS

Cyprinids are primarily freshwater fishes, which means they evolved in a freshwater environment (as opposed to the sea). They are unable to withstand severe salinity. Although a few of these species can survive in brackish water, they are unable to breed there. They are common in the fish fauna of Asia, Europe, and North America, but they are not native to South America or Australasia. A few of these species, however, have now been imported to all parts of the world. For example, the common carp is the most extensively spread alien fish species, with records dating back over 5,000 years. Their prominence in aquaculture is due in part to their ability to feed down in the food chain. Furthermore, many species can withstand high temperatures and low oxygen levels, allowing them to thrive in hostile conditions where other species would perish. This adds to their appeal in aquaculture, where cyprinids account for the majority of global production.

Common carp, tawes (*Barbodes gonionotus*), mrigal (*Cirrhinus mrigala*), big-head (*Aristichthys nobilis*), rohu (*Labea rohu*), grass carp (*Ctenopharyngodon idella*) and silver carp (*Hypophthalmichthys molitrix*), are examples of widely grown species. The others, with the exception of tawes, are all Mekong Basin imported species. There's no reason, however, to believe that native species are inferior. *Cirrhinus microlepis*, the seven-Iine barb (*Probarbus julieni*), *Hypsibarbus suvattii*, and other indigenous species are already being promoted by the MAC Fisheries Program.

Cyprinids are generally gregarious animals that form schools at some point during their lives. Many Mekong cyprinids migrate in large numbers, especially during the dry season. Around December and January, for example, massive quantities of fish migrate upstream from Cambodia through the Khone Falls in Southern Lao PDR. Some species, like the seven-line barb, move hundreds of kilometers into northern Thailand and the Lao People's Democratic Republic, and maybe beyond. Cyprinids are typically fecund since their eggs are tiny. Substrate spawning is the norm, with the eggs typically sticking to rocks or plants. Normally, all of the eggs are released at the same time. It's unusual for parents to look after their offspring's eggs or young.

In the realm of fishing, cyprinids are extremely valuable. Over 4 million tonnes are manufactured or landed worldwide, according to official FAG statistics. Because a huge amount of landings and production is either not documented or is included in statistics under "miscellaneous" species, this figure is significantly understated. Only maritime species such as anchovy, mackerel, and pilchard outnumber one single species, the silver carp, in terms of total production. Most of the world's cyprinid produce is ostensibly derived from "aquaculture," but the statistics are totally controlled by China, where culture-enhanced fish stocks account for a major share of the total (e.g. stocked reservoir fisheries). Cyprinids are without a doubt the most important group in the Mekong. Many Mekong species play an essential role in the riparian countries' fisheries. Some species' migratory behaviors make them ideal targets for fishermen, and a considerable portion of the fisheries is devoted to them (Llorente & Luna, 2016).

It's also feasible that smaller species are just as vital as, if not more essential than, larger ones. The riel (*Henicorhynchus* spp.) accounts for two-thirds of the dai catch in the Tonle Sap River in Cambodia when migrating (December-February). Many additional aquarium fish species are popular and economically useful due to their small size, gorgeous coloration, omnivorous diet, lack of teeth, and resilient temperament. In the Mekong, the potential of the latter resource has yet to be completely realized. For sport fishermen, several of the larger cyprinids are among the most popular targets. For their fighting character, endurance, and difficulty to catch by hook and line, certain species of Tor, for example, are held in greater regard than salmon. With wild salmon in certain European rivers now value more than US$ 25,000 per pound (in terms of asset value), this issue could provide additional justifications for preserving the Mekong fishery.

Carps, barbels, and other cyprinids (or carps) were cultivated in 92 countries in 2017, with a production capacity of 28 million metric tonnes (USD 61 billion), contributing for a quarter of global aquaculture in terms of both quantity and value. India and China are the two most important carp-farming nations, contributing for 71% and 14% of global carp fish production volume, correspondingly, in 2017. According to FAO figures, 38 species of carp were farmed globally in 2017;9 they are predominantly staple fish for household consumption. Other carp species include Carassius spp. (for instance Crucian Carp, *Carassius carassius*); Catla (*Catla catla*); and Labeo Roho (*Labeo rohita*), among the four species of Carp (Grass Carp, Common Carp, Silver Carp) and Bighead Carp).

2.4.1 Catfish vs Tilapia

Catfish and Tilapias are two other freshwater basic fish that make the top 10 rankings for both number and value. In 2017, the ISSCAAP group "Tilapias and other cichlids" (tilapias in short) produced 5.9 million tonnes (USD 11 billion) of world aquaculture tonnage (or 4.42 percent in terms of revenue), while the 5.5 million tonnes (USD 10.6 billion) of cultivated catfishes (i.e. freshwater marine animals of the order Siluriformes) produced 4.9 million tonnes (USD 10.6 billion). Tilapias and catfishes, for example, have a smaller value share in the global aquaculture than their corresponding quantity shares, indicating their low value. Tilapias and catfishes are large international seafood products that are progressively recognized as alternatives for traditional whitefish products, despite the fact that they are predominantly produced for domestic consumption in many nations.

Tilapias were the most frequent aquaculture species group grown in 127 countries in 2017, according to FAO figures. China (27 percent of world tilapia aquaculture tonnage in 2017), Indonesia (22 percent), Vietnam (4.6 percent), the Philippines (4.6 percent), Brazil (4.9 percent), Bangladesh (5.7 percent), and Egypt (16 percent), are the major tilapia growing countries (4.0 percent). While the FAO reports that 18 kinds of tilapia were farmed worldwide in 2017, the Nile tilapia is the most common. Indeed, the 4.1 million tonnes of Nile tilapia production reported by the FAO is understated since the 1.3 million tonnes of Nile tilapia production in 2017 is largely Nile tilapia. According to FAO figures, 27 catfish species were farmed in 86 countries in 2017, with Pangasius catfish accounting for 51% of global catfish aquaculture tonnage and Clarias catfish accounting for the rest (24 percent). Vietnam (24 percent of world tonnage), Indonesia (23%), China (20%), India (9.9%), Bangladesh (9.4%), Nigeria (3.5%), and the United States of America (3.5%) are the major catfish production countries (2.7 %) (Liu et al., 2015).

The Nile tilapia is the most widely farmed tilapia breed in the world, accounting for around 75% of all cultivated tilapia. Tilapia is a popular aquaculture fish because it is very adaptable and can be farmed in a number of aquaculture conditions, including brackish and saltwater, as well as pond and cage systems. Tilapia is an omnivore fish that eats both animal and vegetable-based diets, as well as algae and plankton. Tilapia can deal with low feed quality, rendering them a good candidate for food with a reduced nutritional content than would be required for other aquaculture species. In the grow-out stage, most farmers choose to keep only male tilapia. Male tilapia have proven to be more profitable due to their larger size and efficiency in terms of time and energy. Due to breeding, female tilapia lose a lot of energy and time. Because male and female tilapia are rivals, most farmers try to keep only male tilapia in their ponds. To create only male tilapia progeny, various procedures are currently being used. The following are a few of the genetic breakthroughs in this field.

2.4.2 YY Method

Males with the YY chromosome are produced using this method. To ensure that all XY females are created, the first-generation fry are given estrogen. These are then mated with males who have been identified as XY. After that, this generation is mated with XX females. It can then be established which breeder fish are YY'supermales,' as they will only create XY male progeny as a result of this. After that, these fish can be retained as YY breeding males.

2.4.3 Methyl-testosterone Technique

This procedure involves supplementing the juvenile's food with methyl-testosterone. This successfully transforms females into males while males remain unaffected. The drug leaves no remnant in the fish at harvest because it dissolves after 30 days, yet this method still raises ethical and safety concerns. The fish's liver has been reported to be damaged as a result of this procedure. Other strategies, such as the YY method, have been developed to try to limit the use of MT.

It is advisable to stock one male to three females in a hatchery for optimal results. The eggs should be removed to avoid the females from incubating their eggs orally. This gives the female more control over the hatching process and permits her to generate the next batch of eggs. For large numbers, pond or hapa rearing can be used, while for lower numbers, a spawning tank can be used. Jar hatching has also been discovered to be an effective method of hatching huge numbers of eggs with minimal supervision. Tray hatching is a more effective way for establishing the optimum health conditions and having the most control. However, this has the disadvantage of being quite labor-intensive and thus not the most cost-effective procedure.

Fry are gathered and relocated to grow-out ponds once they have hatched. Tilapia, unlike other fish, are more resistant to low water conditions. The dissolved oxygen level should be around 4mg/l, according to best practice. Low oxygen levels in water can cause mass fish deaths, but tilapia can survive at dissolved oxygen levels as low as 2 mg/l. However, it is not recommended to let the quality drop that low because it will affect the fish's health, quality and growth. Tilapia can withstand a wide range of pH, however, a pH above 6.5 promotes algae growth, which provides food for the fish. The fish prefer temperatures of roughly 25 degrees Celsius. Tilapia can be kept at up to 60 kg/m3 density (depending on water and management conditions). High stocking densities, on the other hand, may have a negative impact on growth and health. Low stocking densities, on the other hand, frequently result in fish antagonism (Lafont et al., 2019).

The majority of catfish are raised in earthen ponds, which are typically 5 to 10 acres in size, with cages and raceways producing only a small percentage of the total. The split-pond system, a new production system created at Mississippi State University, has been proved to increase yields, and farmers are beginning to implement it on their fields. The majority of catfish are produced in many batches. Small catfish (fingerlings) are often introduced in the spring, with market-sized fish being removed by seining on a regular basis. Wastes are eliminated through natural processes within the ponds, therefore ponds can be kept for years without being drained.

Fingerling catfish are supplied and raised until nearly all of the catfish have achieved the size of the market in single-batch production, after which the whole pond of fish is taken at once and the pond is re-stocked. The ground feed materials are blended and pushed through an extruder in the manufacturing of catfish feed pellets, which float. Farmers utilize a truck or trailer-mounted hopper with a blower to spray feed pellets over the water surface to feed fish. As a result, the fish feeding reaction can be easily monitored to ensure that the fish are fed adequately but not excessively. During the growth period, dissolved oxygen levels in catfish ponds decrease at night, so growers use aerators (usually electric paddle-wheels) to boost oxygen levels in the water.

Catfish output in the United States has decreased in recent years, owing to competition from Pangasius species (all catfish are classified as Siluriformes) imported from Vietnam and sold under the names Swai, Tra, and Basa. Chinese and Indonesian channel catfish fillets are also imported. In some markets, imported tilapia goods appear to compete with catfish. High feed prices pushed up production expenses, especially as feed is a major operating expense. Catfish farming has become profitable again, thanks to recent increases in catfish prices offered to farmers, and farmers are ramping up production to meet a demand for farm-raised catfish in the United States.

2.4.4 Brown seaweed vs Red Seaweed

Seaweed farming in the United States has exploded in recent years, with dozens of farms operating in New England, the Pacific Northwest, and Alaska, as well as more in the works. Seaweeds such as dulse, bull kelp, ribbon kelp, and sugar kelp are grown by farmers and utilized in sashimi, salsas, sauce, salad, seasoning, and other foodstuffs. Cosmetics, livestock feed, and manure are all made with farmed seaweed. Seaweed farms thrive by absorbing nutrients and carbon dioxide. The farms have the potential to improve water quality and mitigate the consequences of ocean acidification in the surrounding area.

Seaweed provides a chance for shellfish and fin-fish growers, commercial fishers, and first-time farmers to diversify their farming operations or start a new business. Seaweeds take advantage of the entire water column. This implies growers may grow seaweed vertically, or in three dimensions, and reap big harvests from a tiny area. During the winter, most seaweeds grow on longlines suspended 4–8 feet below the surface. Before being harvested in the spring, their blades will have grown to a length of 10 feet or more. The fastest-growing aquaculture industry is seaweed cultivation. In 2019, Alaskan farmers produced about 112,000 pounds of sugar, ribbon, and bull kelp, for instance. This represents a 200 percent growth over the state's first commercial harvest, which occurred in 2017. Off the coast of southeastern Alaska sits the world's largest kelp farm (Kaminski et al., 2018).

However, the industry is still in its infancy. NOAA and its partners are looking into the regulations for siting, as well as the architecture, equipment, and best management techniques. This effort will aid in the development of more efficient permitting, allowing seaweed farming to grow while remaining environmentally and economically viable.

Seaweeds are masters at soaking up carbon dioxide and converting it into energy. The potential of eelgrass, mangroves, and salt marshes to store carbon is well documented. However, based on biomass, seaweeds remove more greenhouse gas from the ocean than the other three combined. As a result, seaweed farms may be able to help mitigate the local effects of ocean acidification. Seaweeds consume nitrogen and phosphorus as well. When these nutrients are present in significant numbers, they generate algal blooms, which deplete oxygen in the water as they disintegrate. The dead zones that emerge in the Gulf of Mexico and the Chesapeake Bay are caused by excess nitrogen and phosphorus from stormwater runoff and point sources. Seaweed farms can aid in the reduction of nutrient levels in local bodies of water.

Kelp forests are a vital coastal environment found all over the world in temperate climates. Bull kelp has dropped dramatically in Puget Sound. NOAA is collaborating with local partners to restore these enormous, fast-growing kelps using restoration aquaculture techniques. For commercially significant fish and a diverse range of marine species, seaweeds provide food and refuge habitats. Seaweeds can also be used as a sponge, absorbing whatever is in the water. Heavy metals and other coastal pollution could be reduced by cultivating these marine plants. The seaweeds raised for this purpose, unlike in commercial seaweed cultivation, would not be eaten. Rather, communities and other stakeholders could create seaweed farms in contaminated places only to improve the health of coastal ecosystems.

Red and brown seaweeds (two ISSCAAP groups) are the number 2 and 3 farmed species individuals in terms of amount, but they have dropped out of the top 10 value list; brown seaweeds have dropped to number 11 and red seaweeds have dropped to number 13. According to FAO figures, eight red seaweed species were farmed in 32 countries in 2017, with China and Indonesia responsible for 28 percent and 56 percent of global output tonnage, correspondingly. According to FAO statistics, there were eight brown seaweed varieties cultivated in 12 countries in 2017, with China being the largest producer (responsible for 87 percent of global output tonnage), trailed by the Republic of Korea (8.9%) and the Democratic People's Republic of Korea (4.0 percent). In addition to kelps, which are used for both human and industrial purposes, Wakame (*Undaria pinnatifida*) is a prominent brown seaweed used primarily for human consumption, with 2.3 million tonnes produced in 2017 (2.1 percent of total world output) ranking it 16th among all species.

2.4.5 Clams vs Oysters

According to a Rutgers University–New Brunswick study, eastern oysters and three types of clams may be cultivated together and thrive, potentially increasing shellfish grower profits. Despite the fact that diverse groupings of species outperform single-species groups, most bivalve farms in the United States and around the world cultivate monocultures, according to a study published in the journal Marine Ecology Progress Series. By allowing shellfish growers to more readily manage market pressures if the price of each individual product fluctuates, farming numerous species together can help farm operations maintain economic sustainability and boost profitability.

Mollusk farming, like scallops, ousters, and clams, gives billions of dollars to the global economy each year. In 2016, more than 47 million pounds of farm-raised oyster, clam, and mussel meat value more than $340 million were collected in the United States, according to the report. The research was conducted in a lab environment at Rutgers' New Jersey Aquaculture Innovation Center in North Cape May, New Jersey, to see if it was possible to farm different bivalve species in close proximity to one another (Jothiswaran et al., 2020). The study looked at the filtration rate, growth, and longevity of four ecologically and economically significant bivalve species native to the northeastern United States under farm circumstances. The Atlantic surf-clam (*Spisula solidissima*), soft-shell clam (*Mercenaria mercenaria*), the hard clam (*Mercenaria mercenaria*), and the Eastern oyster (*Crassostrea virginica*) are the four species (*Mya arenaria*).

When given saltwater containing naturally produced algal particles, subgroups with all four species eliminated substantially more particulate matter than monocultures. This indicates that each species prefers to filter a certain type of algal food particle. The researchers also discovered that none of the four species had significantly different growth or survival rates, implying that if food is not a concern, these bivalves might be raised with each other without outcompeting one another. The advantages of diversifying crops on shellfish farms are demonstrated in this study. Crop diversity protects aquaculture growers from individual crop failures caused by disease, predation, or changing environmental circumstances. The practicality of bivalve polyculture should be evaluated on commercial bivalve farms in future studies.

2.4.6 Marine Prawns and Shrimp

Commercial marine fish production began in the 1970s, and output quickly increased to meet the demands of the US, Japanese, and European Countries markets. In 1991, global cultivated shrimp production totaled more than 2.1 million tonnes, with a market value of approximately US$9 billion. In Asia, about 30% of farmed shrimp is generated, primarily in Indonesia and China. The remaining 54.1 percent is primarily produced in Latin America, with Mexico, Ecuador, and Brazil, leading the way. Indonesia is the world's greatest exporter.

Shrimp farming has evolved from a small-scale, regional sector in Southeast Asia to a global industry. Technological advancements have allowed shrimp to be grown at increasing larger densities, and brood-stock is now sent all over the world. Almost all farmed shrimp belong to the Penaeidae family, with only two species accounting for 32.14 percent of all farmed shrimp: *Litopenaeus vannamei* (Pacific white shrimp) and *Penaeus monodon* (giant tiger prawn). These industrial monocultures are highly susceptible to illnesses, which have resulted in the extinction of farm shrimp populations in some regions. Increasing ecological issues, disease outbreaks, and pressure and criticism from NGOs and consumer countries prompted changes in the business in the late 1990s, as well as typically better government oversight. In 1999, a program including governmental entities, industry leaders, and environmental organizations was launched with the goal of creating and supporting more sustainable farming practices.

Freshwater prawn farming has a lot in common with marine shrimp farming, and it has a lot of the same issues. The developmental lifetime of the principal species introduces unique issues (the giant river prawn, *Macrobrachium rosenbergii*). In 2010, the global yearly production of freshwater prawns was at 670,000 tons, with China accounting for 615,000 tons (92 percent).

2.4.7 Smelts, trouts, and salmons

Salmonid aquaculture involves the growing and harvesting of salmonids in regulated environments for recreational and commercial purposes. Salmonids (especially rainbow trouts and salmon), carp, and tilapia are the 3 most significant farmed fish species. The Atlantic salmon is the most common commercially farmed salmonid. Through the National Fish Hatchery System in the United States, rainbow trout and Chinook salmon are the most extensively farmed salmonids for leisure and subsistence fishing. Brown trout are the most widely produced fish for recreational restocking in Europe. Tilapia, sea bass, catfish, and bream are examples of non-salmonid fish that are commonly farmed.

Salmonid aquaculture was value US$10.7 billion worldwide in 2007. Between 1982 and 2007, salmonid aquaculture productivity increased by more than tenfold. Scotland, Chile, Norway, and Canada were the top salmonid producers in 2012. The ecological and physiological effects of industrial salmonid aquaculture are hotly debated. The effects on wild salmon and other marine animals are particularly concerning. This debate is part of a larger economic battle for market share and price between Alaska commercial salmonid fishermen and the rapidly developing salmonid aquaculture business (Jensen et al., 2014).

Commercial fishing tactics are used to capture wild salmonids from their natural habitats. Russian, Japanese, and North American fisheries catch the majority of wild salmonids. There are changes in wild and farmed salmonid production during a 25-year period, as reported by the FAO. Alaska, Japan, and Russia are all involved in large-scale hatchery-based stock improvement operations that are essentially ocean farming. For FAO and marketing purposes, the resulting fish hatchery fish are classified as “wild”.

2.4.8 Crayfish

Australia is among the world’s wealthiest freshwater crayfish archives. Every one of our species belongs in South America, Madagascar, New Zealand, New Guinea, and Australia, to the Parastacidae family. Our freshwater crayfish vary from the biggest to the smallest. The Tasmanian giant crayfish

(*Astacopis goulardii*) grows to 6.3 kg, while the Murray River is 2.7 kg. Their slow growth and low meat yields , in general, have considered these species and spiny crayfish to have little aquacultural potential.

Three Cherax species have been identified as having aquaculture potential. The yabby (Cherax destructor) is endemic to much of Queensland, South Australia, Victoria, and the red-claw (*Cherax quadricarinatus*) is native to northern Queensland and the Northern Territory; and New South Wales; the marron (*Cherax tenuimanus*) is native to south west Western Australia. The potential for semi-intensive aquaculture of the two Cherax species inside New South Wales red-claw and yabby is compared in this report. Claws are tiny in both species. The meat yield (meat to body weight ratio) of red-claw is 22-25 percent; the meat yield of yabbies is somewhat lower.

The domestic market's size is unknown. For years, yabbies from the wild fishery have been sold in tiny quantities. By selling live crayfish of a specific size to restaurants, far greater prices (over US$20/kg) have been obtained for both species. Farmers are currently paid a variety of prices for live crayfish, which vary depending on size, buyer, and season. The NSW DPI aquaculture production report has average price information. Large retail stores, such as the Sydney Fish Market, sell live crayfish on a daily basis.

It's difficult to foresee the size of any export market until supplies can be sustained. Crayfish has a market in Europe since they are considered a delicacy and local numbers have been depleted by the "crayfish epidemic." However, in several European nations (such as Sweden), the market is particularly seasonal. *Cherax* spp. have been accepted in the marketplace based on trial shipments. Some products have been sent to Asia and have been well accepted, so there is potential there as well. The volume and consistency of supply will determine the success of future export trade.

Crayfish aquaculture is a steady sector in NSW, however, there is still a lot of room for growth. Because of its natural occurrence and endurance to a wide range of circumstances and climates, the yabby is the most ideal species for NSW. More research is needed to discover yabby production methods in semi-intensive purpose-built pond environments. Crayfishes are ranked #15 by quantity and #6 by value in the ISSCAAP group "Freshwater crustaceans," which comprises freshwater decapod crustaceans of the superfamilies Astacoidea and Parastacoidea. While the FAO reports that seven crayfish species were farmed in 2017, red swamp crawfish account for nearly all of the world's crayfish aquaculture production.

2.4.9 Crab

Crab enthusiasts need not worry, according to scientists, because pond-raised crabs look and taste exactly like their ocean-raised counterparts. Since the native blue crab population in North Carolina has been at historic lows as of 2000, Dave Eggleston, director of NC State's Center for Marine Sciences and Technology (CMAST) and scholar of marine, earth, and atmospheric sciences, has been researching various methods to help the population recover. He came up with a solution that benefits farmers trying to diversify their crops while simultaneously reducing strain on existing crab populations: growing blue crabs in irrigation ponds on farms.

They began by catching small crabs in the wild and storing them in ponds full of bass and bluegill predators, which showed a survival rate of 12 percent. They then worked together in the NC State's Vernon James Research and Extension Center in Plymouth, North Carolina, with the University of Maryland's Marine Biotechnology Center that has the knowhow to grow hatchery-reared blue crusts, and stored them in freshwater experimental aquaculture tubs, where crabs demonstrated some the highest growth rate records (Holmen et al., 2018).

In reaction to the downturn in the tobacco market, Eggleston discovered that many farmers in Eastern North Carolina were attempting to diversify their crop offers. He wondered if the irrigation ponds on many of these farms could be used to raise crabs, providing farmers another crop to focus on and relieving strain on the coastal crab population. Crabs can handle a saline level of only 0.3 parts per thousand, which is similar to the level seen in coastal tap water, according to Eggleston and his colleagues. They went on to research the ideal circumstances for producing crab, including population density, feeding rations, and pond habitat structure.

Eggleston and Ray Harris, NC's State Director for Carteret County Cooperatives in July of last year, had the permission to finish a large-scale test when they packed a ten-acre lake of 40,000 hatchery crabs and a smaller 4,000 crab pond. The crabs take about 105 days to mature and the results are promising. Because of the quick growth rate of these pond-raised crabs, Eggleston estimates that a farm may generate two to three catches in a single year, as crabs do not perform well in freshwater during the winter months. He claims that if you put 50,000 hatchery-raised crabs in a 2.5-acre pond, you may expect to collect roughly 20% of them, or 10,000 fully developed crabs. The crabs would fetch US$30,000 per harvest at US$3 per crab.

2.4.10 Perch-like Fish

Perca flavenscens, or yellow perch, is a popular sportfish found across the Great Lakes region and prized for its high-quality, firm, white, flaky flesh (Hinshaw 2006). During the 1950s and 1960s, a commercial yellow perch fishery produced as much as 37 million pounds of fish, but it nearly collapsed shortly after (Scott & Crossman, 1973). Commercial yellow perch harvest was banned in response, with the exception of Lake Erie and Lake Michigan's Green Bay. The commercial yellow perch catch from the Great Lakes was expected to be 8.33 million pounds in 2008.

Despite a significant drop in commercial harvests, yellow perch demand as a food fish has remained high for more than 30 years (Riepe, 1998). Researchers at the University of Wisconsin-Madison became interested in culturing/farming the species as early as the 1970s (Hart et al., 2006). Yellow perch has qualities that made them a good choice for commercial aquaculture, according to preliminary research:

- A pelleted fish food was quickly accepted.
- With little impact on growth, they dealt with crowding, handling, and water quality degradation.
- At high numbers, they did not engage in cannibalism.
- The flavor of fish raised in ponds and tanks was shown to be unaffected (Hart et al., 2006).

The culture of yellow perch is fraught with difficulties. In many culture species, a lack of understanding of nutritional requirements has been a problem. Yellow perch have yet to be given a commercially available specialist diet, hence they are normally fed rainbow trout diets (Hinshaw., 2006). Yellow perch also reach sexual maturity before they reach market size. Rather than producing flesh, little perch focus their efforts on developing reproductive organs and spawning. There has been research towards developing sterile fish and all-female production (Hart et al., 2006).

The last, and maybe most serious, issue is the scarcity of fingerlings for grow-out. A sustained and consistent supply of fingerlings is required for an economically viable industry. Yellow perch spawning protocols have yet to be created in a consistent manner. Furthermore, no designed or artificial yellow perch fry first feed diets have been produced. Even with known pond fertilization procedures, pond productivity cycles in the midwestern United States, where the majority of yellow perch production occurs, remain varied from year to year.

As a result, the timing of zooplankton blooms, which is crucial for fry survival in outdoor systems, might be significantly out of sync with the timing of the first feeding requirements. As a result, intensive indoor fry production appears to provide superior management. Indoor first feeding procedures, on the other hand, are currently expensive and time demanding, with very varied survival rates. For first feeding, the majority of commercial yellow perch facilities in the United States use outdoor production techniques, and for fingerling feed training, they use either outdoor or outdoor/indoor hybrid methods.

Freshwater perch-like fish, a subset of the ISSCAAP category "Miscellaneous freshwater fishes" that contains freshwater fishes of the superfamily Percoidea, was rated #18 in terms of amount (0.8 million tonnes; 0.7 percent of all species) and #9 in terms of value (USD 7.1 billion; 2.85 percent of the world value of production). This is a high-value species category, as seen by the substantially larger value share (Hall et al., 2021).

According to FAO figures, nine freshwater perch-like fishes were farmed in 30 countries around the world in 2017, with a total production of 0.8 million tonnes. The largest species in this group is the largemouth black bass (*Micropterus salmoides*), which accounts for 56 percent of total production. While the FAO reports six nations cultivating the species in 2017, China accounted for 99.7% of global production in 2017. According to FAO figures, China was the only producer of the species in 2017. Mandarin fish (*Siniperca chuatsi*) accounted for 41% of global aquaculture production of freshwater perch-like fishes in 2017.

2.4.11 Milkfish

Around four or six centuries ago, milkfish production began in Indonesia, Taiwan, China and the Philippines. Cultural methods are continually improved in a variety of enclosures.

Since the 1970s, significant investment in infrastructure, research, loans and training in support of the milkfish industry has been made in the Philippines, as well as the Taiwan Province of China, Hawaii and Indonesia. For instance, in 1973, a special task was set up in Iloilo, the Philippines, to analyze and solve problems of milkfish aquaculture in the Center for Southeast Asian Fisheries Development (SEAFDEC) in Iloilo,

Philippines. From the mid-1970s until the present, government agencies and fishery institutions were also involved in a national drive to expand milkfish farming.

Research and development was carried out on agricultural systems, breeding and frying technology. No attempts have been made to genetically enhance, but fry translocation and trading between Indonesia, China's Taiwan Province and the Philippines have been shown. Recent unconfirmed reports have shown that milk is now cultivated in the South Pacific Islands and in Singapore as a tuna bait to finger or juvenile size. The dairy farming industry had previously been traditional and the production in captivity of sexually mature reproduced fish was a little emphasis. The traditional dairy industry was completely dependent on an annual rehabilitation of farm ponds with fingerlings from wild fries. This has caused changes in fry availability in the industry on a regional, seasonal and annual basis. In general, these variations can be unpredictable and over short periods of time quite large (Gjedrem, 2012). As a result, the international milkfish industry's main challenge was to devise a method of producing a consistent, adequate, and high-quality supply of milkfish fry that was not subject to massive, unforeseen changes in time and place. There has been considerable progress over the last ten years, especially with regard to the spread of milkfish and mass fried production by personal hatchery, research institutions and government agencies. Rather than relying on wild fried fish, milkfish farms in the Philippines, China's and Indonesia's Taiwan provinces now get most of their fried fried hatcheries, hatcheries due to the shortage of wild fries. Milkfish, the world's #13 species category in terms of amount, is ranked #9 in terms of quantity and #10 in terms of value in world aquaculture, except China. It has two large producers Indonesia (73 percent of 1,7 million tons of world milkfish aquaculture produced in 2017) and the Philippines, a diadromatic fish (the only live fish in the Chanidae family) that are primarily grown and consumed in Southeast Asia and the Philippines (24 percent).

2.4.12 Mussels

Aquaculture, or aquatic farming, is used to raise blue mussels since many organisms, particularly humans, like eating them. According to the Maine Department of Marine Resources, more than 650,000 pounds of blue mussels were cultivated and harvested in 2012.

Young mussels can be grown in the wild or on ropes immersed in culture tanks, in which they are sheltered from predation and thunderstorms. The mussels are put into ocean pens to grow after they reach a particular size.

However, practitioners frequently face challenges in increasing the quantity, size, and overall health of their mussels. To boost their produce, they, like many other farmers, turn to science and engineering. Beal is working with a team of NSF-funded researchers from the University of Maine at Machias and the Downeast Institute to figure out what growing conditions and methods will reliably produce healthy and abundant blue mussels.

"Our objective in the nursery is to create methods for producing consistent amounts of seed-size mussel juveniles," Beal said. "At the moment, mussel farmers rely on wild settlement, which is inconsistent from year to year and location to location." The annual harvest of cultivated blue mussels in Maine can range from hundreds of thousands to millions of pounds. Mussels in their infancy go through various stages of development. Mussel larvae stick to an aquatic surface such as a rope after swimming during the first few weeks of their lives. They use byssus threads, which are flexible protein strands, to attach themselves.

"A small range of seawater temperature mixed with significantly higher salinities resulted in healthy, active juvenile," Beal explained, "and varied phytoplankton diets provided to aquatic larvae impact their capacity to settle efficiently onto surfaces like rope." Beal's team hopes to apply what they've learned about blue mussel growth to improve the number of larvae that can be attached to aquaculture rope and how well they do so. They're now doing field research to see how different stocking densities affect mussel development and survival. The experts are also looking at when the newborn mussels should be introduced to ocean enclosures and where they should grow the best (Gopakumar, 2003).

The researchers and their aquaculture partner, New DHC, want to boost economic prospects for sustainably cultivated blue mussels with a greater understanding of their cultivation. "A continuous seed supply will also enable aquaculturists to establish business plans that more realistically forecast their annual production," Beal stated. The NSF Partners for Development program, which encourages regional innovation based on scientific and engineering breakthroughs, is supporting the cooperation. "His life's work is of huge opportunity relevance to the business and the ethos of the region," NSF program director Sally Nerlove said of Beal. "At the same time, his successes reflect the history of the NSF Partnerships for Development program."

2.4.13 Others

The four mainly intensively cultivated species of rainbow trout (cultivated in both freshwater and sea) and sea fish Atlantic salmon, dormitorium seabeam and European sea bases constitute 80 percent of the cultivated fish production in the EU. In Norway, 98% of the production of farmed fish is Atlantic salmon and rainbow trout. Fish meal and/or fish oil feed are essential for rainbow trout and marine fish. They account for a small percentage of total farmed fish output tonnage, but they absorb a significant share of the global fish oil and fish meal required by aquaculture. The intense cultivation of these fish, which confines enormous numbers of animals in a limited area, results in a variety of major pricesafety issues.

The EU and the US are the top suppliers of Pangasius catfish, which are farmed rigorously in Vietnam. Half of the output is exported. In the recent decade, Vietnamese Pangasius production has expanded 10-fold, reaching 1.1 million tonnes in 2010 (78 percent of global farmed Pangasius production), of which 0.66 million tonnes were exported. A rapid rate of water interchange and the capacity of Pangasius to breathe atmospheric oxygen allow them to withstand low amounts of dissolved oxygen and highly dirty water, allowing for extremely high stocking densities (Frankic & Hershner, 2003).

Intensive monocultures rearing Pangasius in pools (with water exchanges and pumpings for nearby river tributaries) and net cages and net pens in major river tributaries of the Mekong River Delta replace small pool systems. Small pool systems are replaced. Intensive ponds and net stalks of monoculture are generally stored at 40 to 60 fish each meter, and even sometimes greater. Net cages typically have stocking densities of 100-150 fish per square meter.

A study has indicated that Pangasius in Vietnam are produced in deep (4m deep) pools at stock densities of 44 fish per m2. The Pangasius is 1kg in terms of market size. Death rates are reported to be between 20 and 25 percent for these fish. In Bangladesh, stocking numbers for this species, according to the same study in 1m deep basins, are substantially lower at 6 fish per m2 and the mortality rate is significantly lower at 10%. The majority of farms in Bangladesh also stock about 10 to 20 percent filter carps that eat algal blooms generated in fertile ponds, while Pangasius in Vietnam is grown in monocultures (made so by the feed inputs).

The carp needs dissolved oxygen and is much less tolerable than Pangasius for poor water quality, implying that quality of water remains at relatively high levels. According to the study's researchers, Bangladesh also saw decreased use of antibiotics for improved levels of fish pricesafety (in terms of growth performance, water management, and mortality rates). In Bangladesh and Vietnam both farmed and commercial feeds, which contain wild fish, are frequently used. Grass carp that are carefully maintained in cages are supplied with other carp (generally as the dominant species), and the cages are typically 60 m2 in size with a depth of 2-2.5 m. The seaweed and groundwater pelleted or other commercial food are feed to the fish. 10 to 20 grass carp per cubic meter are stocked. Wuchang bream are also stocked at a rate of 35 to 50 fish per cubic meter. As 'cage cleaners,' a limited number of silver and bighead carp (1 percent of the total fish) are also stocked (to eat and control algae).

Chapter 3

Raising Ornamental Fish Extensively

CONTENTS

3.1 INTRODUCTION

Over the years, there has been increasing demand for fish, which has paved way for the worldwide trade of ornamental fish all around the world. Fish farming is normally termed as aquaculture and normally involves the rearing of fish in most cases for food purposes or rather for artificial enclosures. Some people are rearing fish for aesthetic value and it is now a lucrative business and some companies have even patented some of the fish. It is commonly a sub-type of aquaculture as it involves controlled cultivation and harvesting of animals that are found in the aquatic environment. The most common types of fish that are commonly reared in fish farms include; carp, catfish, tilapia and salmon. In the case of ornamental fish, the species are different.

Figure 3.1: Fish species.

In most cases, ornamental fish are those that are kept because of the fact that they are attractive. It was over 1000 years ago that the rearing of ornamental fish started such as the raisin of goldfish in china. In the current society, many other species have been recognized and are being reared for their distinct marking. This includes the koi carp. In the United Kingdom, it is estimated that 1 in 10 households have pet fish, which is even topped by the United States where more than 20% of the households have at least a single pet fish. Majority of the individuals that keep ornamental fish keep the gold fish given the fact that they are cheaper. In an aquarium, approximately 25 million fish are kept while in the United Kingdom; 20 million are kept in the ponds alone. Around the world, there are 4000 species of fish that are kept as pets or in public displays. There are billions of money stated as the net value of ornamental fish traded annually for retail value, 20 dollars is the price for a single ornamental fish (Fore et al., 2018).

The types of fish that are kept, as pets are divers ranging from a number of them that are from natural habitats and for that reason all of the pricesafety troubles already referred to in retaining wild fish in captivity and in aquaculture that include the decorative fish seem to be dealt with in a way. Many species are collected in the wild and transported for up to a few days from regions consisting of the Amazon and coral reefs. During shipments of wild-caught fish, cases of mortality normally range from 1 to 30% for the reason that fish undergo negative situations and each transportation and acclimation can bring about pressure within the live-fish chain. Mortality rates also can be excessive after delivery especially due to poor handling and storage by the individuals receiving the fish. Generally, 90% of freshwater ornamentals are farmed, while the best 10% are amassed within the wild.

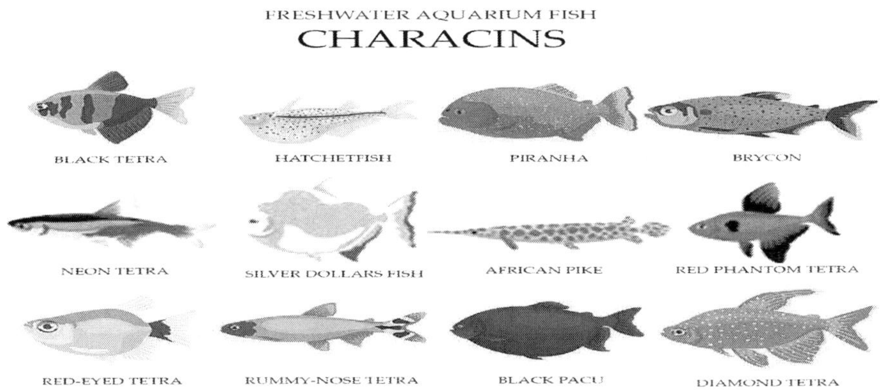

Figure 3.2: Fresh water ornamentals.

Source: Shutterstock

Farmed decorative fish are frequently produced in large-scale centers in, for example, Israel and the Far East, and account possibly for pricesafety worries in times when the fish experience disease and mortality. For example, in most cases, potentially devastating viral outbreaks had been mentioned from numerous of those centers involved in raising ornamental species of fish. For marine species, around 95% are collected in the wild and simply 5% are reared in ponds or aquariums. As such, the tropical marine fish change might additionally have an effect on natural habitat populations and has poor results with the case of conservation and biodiversity. Various severe morphologies had been produced via means of decorative fish breeders consisting of fancy bubble eye goldfish and balloon mollies, which include mollies with stunted frame, form and distended abdomens. All of the above compromise pricesafety because of pressure and unwell fitness, plus some of the hybridized species had been produced that include the parrot cichlid, for the purpose of his or her decorative appeal, and the pricesafety results of those are unknown and as such extra proof is needed.

This is a comparable scenario to pedigree dogs and cat breeding in which positive breeds decided on for developments has brought about fitness problems such as for the case of snout shortening in a few breeds of canines that is known to result in brachycephalic obstructive airway syndrome. Fish are distinctly less expensive pets and, typically speaking, any grown up can buy an aquarium or pond fish understanding not anything such as retaining fish besides water and meals requirements. Taking for instance in evaluation to retaining mammalian pets in view that people are mammals so have a higher understating of mammal husbandry and care. Reputable pet shops frequently enquire approximately on the owner's knowledge and may offer recommendations on tank size, water situations, and some other aspects applicable to the unique species. However, it is not all owners show off such suitable exercise and may bring about pressure and negative pricesafety within the arms of the owners that are inexperienced.

Figure 3.3: Goldfish in a pet shop.

Source: Alamy

It is advised that there needs to be a ban in every country with regard to mutilated pet fish such as the dye-injection, dye-tattooed and those that have gone through cosmetic surgery. Not only should they be banned, restrictions need to be made with regards to certain man-made constraints that are of pricesafety concerns. Such strains include the scoriotic parrot cichlid hybrids, the celestial-eyed goldfish and the balloon mollies that are known to experience swimming problems because of the fact that their body shapes are unnatural. These are some of the aspects that are considered important f the pricesafety of the fish is something that is considered a priority. There are those species of fish that grow to be too large to be able to fit the home aquaria. Such need restrictions in order to help the fish from being held in stressful situations and end up dying prematurely. Such species of fish include the Pangassius, red-tailed catfish and shovel-nosed catfish and as such, when most individuals ignore this aspect, they may even end up being rehomed to a certain public aquarium (Fiorella et al., 2021).

In the case of wild-caught fish that are then transferred a tank at home, they tend to have higher mortality rates. This is because of the stressful conditions that the fish go through in the process of transportation and due to the acclimation stress in fish. Guidelines normally allow for up to three days of survival while transporting fish species such as the common clownfish, a transportation time of more than 24 hours may result in stressful conditions. As such, instead of stressing the fish by exporting them to another country, the importing country needs to reduce such cases by banning the importation of the fish and lowering the time for transportation of the same. In the process, it will improve the fish price safety and further reduce the mortality rates.

In India, the trading of ornamental fish is known to be a common aspect and has actually become very popular in Tripura. The trade of the same is normally confined to the importation of exotic ornamental fishes from other states but within the country. This ensures that the transportation time is reduced thereby lowering the stress levels of the fish and low cases of mortality of the ornamental fish. There are those indigenous ornamental fishes that are available in a number of states that are normally neglected because the majority of the individuals are unaware of the same or the aspects that they need to consider in order to ensure their safety. In most instances, these fish are normally caught before they are able to reach their maturity levels and then sold to the market as common food fishes. The hardship that most individuals experience in the country is the route reason as to why most individuals run to harvest the fish even before they reach maturity and thereby not giving them a chance to see the beauty of the same.

Figure 3.4: Ornamental fish in the river.

Source: Times of India

3.2 CULTURE OF ORNAMENTAL FISH

Differed countries have different cultures when it comes to dealing with ornamental fish. In fact, different countries consider different species as important ornamental fishes. Take India for example. Generally, the exotic fish are known to dominate the Indian marketplace and greater than 250 types had been recorded from the home marketplace. More than 200 species of those freshwater fish are bred in exceptional waters of India with the aid of using the ornamental fish breeders and aquarium hobbyists. The egg layers lay adhesive or non-adhesive eggs at the glass wall or aquarium plants (Erondu & Anyanwu, 2005). Some of the fish parents display parental care, a few smash their eggs, and for this reason, exceptional breeding setups are needed. In contrast, constant bearer fish launch their young ones in batches and are easy about their ability to reproduce in comparison to egg-layers. Common exotic livebearers like guppy, molly, swordtail, platy and egg layers like goldfish, koi carp, tiger barb, Siamese fighting fish, tetra, silver shark, angel, purple-tailed black shark, red finned shark, among others are known to dominate the marketplace. Native ornamental fish are inclusive of honey gourami, rosy barb, zebrafish, glassfish, and loach, among others in India.

Figure 3.5: Honey gourami is a native ornamental fish in India.

Source: Fishkeeping World

Most of the common cultural centers used for decorative fish are cement cisterns, glass aquaria, earthen ponds, and earthen pots, among others. Three to 4 cement cisterns are enough for a small-scale rearing unit such as (3 x 2 x 1m) and are constructed above the floorstage for smooth drainage. All glass aquaria are favored for breeding functions because of the fact that warmers and aerators may be used easily. Even, fish farmers with small earthen tanks can use them for rearing juveniles with the fish that are used for the purpose of food. Marginal farmers even can use huge earthen pots of 1.5m diameter for the rearing of larvae and juveniles. Generally, rainwater is the quality supply of water for ornamental fish lifestyle. If the municipal water supply is in use one needs to be able to aerate it before use for a couple of days for de-chlorination. Tube well water is likewise used at once within the rural areas. The common temperature of the rearing water within the place is 15-28 and also 176 degrees Celsius and ensure that the pH is barely alkaline. Most of the species cultured opt for soft to medium-hard water.

3.3 FOOD AND FEEDING OF ORNAMENTAL FISH EXTENSIVELY

The small-scale farmers do not have enough money distinctly to allow them to purchase readymade packed pellet feed or buying of artemia cyst drums that are costly. However, they have the correctly substituted live feeds. Water fleas, Tubifex or sludge worm, mosquito larvae and chopped earthworm are utilized by the farmers. Different homemade feeds like whole-wheat bread, vegetable peelings and rice also are fed to the fish. However, maximum farms rely on Daphnia, Tubifex worms and mosquito larvae for feeding the ornamental fish. The fish culturists can accumulate Daphnia from the close to through ponds through sieving by doing first-rate mesh within the early morning. Tubifex worms and mosquito larvae are collected from the sewage water channels. Generally, the farmers dispense the feed as much as they can daily. Overfeeding is extra dangerous than underfeeding of the fish because the extra feed destroys the water quality.

Figure 3.6: Feeding ornamental fish extensively.

Source: Algone

Just like the other types of extensive fish, ornamental fish that are reared extensively feed on commercial feeds and not on aquatic plants. Given the fact that extensive fish farming tends to be contaminated by the feeds, it is a requirement that the farmer ensures to feed the ornamental fish just enough food in order to ensure proper aeration and thus preventing suffocation of the fish and reduce the mortality rate. In order to ensure proper ornamental breeding in an extensive manner, there needs to be proper quality management. This is known to be a proper preventative measure given the fact that ornamental fish are very sensitive to both temperature and pH. White spot, mouth fungus, tail and fin rot are some of the common diseases that are faced by ornamental fish. However, some economical chemicals and medicines are easily available and can be used as preventive measures. Some examples of such easily available remedies that can prevent cases of diseases include methylene blue (2.5 g/L) added in aquarium water used for water purification, copper sulfate or potassium permanganate (0.5-1 g/L) used as bath treatment for 1 min as disinfectant and salt (15-30 g/L) used as bath treatment for 30 min as a disinfectant (Evers, 2010).

Culture and breeding of the ornamental species of fishes may be a promising opportunity for lots of human beings in addition to unemployed youths. It calls for the little area and much less preliminary funding than different types of aquaculture. For decorative fish farming, it would be best to have a clean knowledge of the conduct and biology of the fishes. It may be practiced even in city regions with little alteration of outdoor or roof of a house. As much less workforce is needed, the young ones or even the elders can run small domestic aquarium gadgets and enhance their social and financial status.

There are, however, some challenges that human beings may encounter when rearing ornamental fish extensively. Such include the fact that humans may be exposed to such as a result of being in direct contact with the fish or the water that they are being reared in. Such diseases are defined as zoonotic diseases. Most of the zoonotic diseases that are connected to the ornamental fish are because of the microbes that get into contact with the open wounds. Alternatively, the puncture wounds that such individuals experience whenever they are handling such equipment. In addition to this, those human beings that have a compromised immune system may be predisposed to what is described as opportunistic infections.

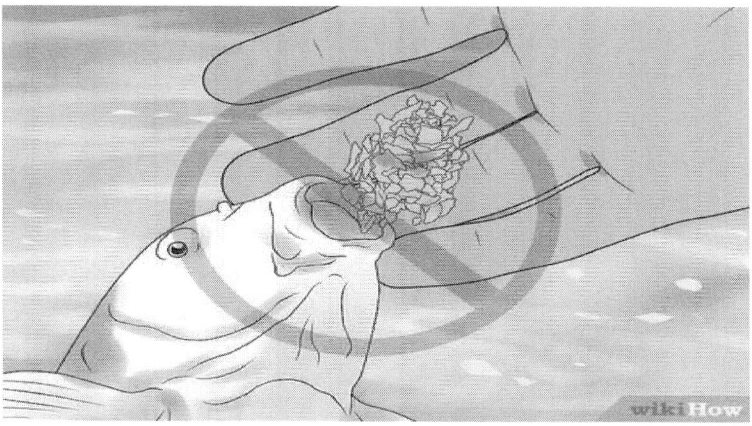

Figure 3.7: Handling fish without gloves is risky.

Source: WikiHow

In most cases, the fish-borne diseases normally result in mild episodes of diseases. Some of the opportunistic pathogens that are associated with human beings include, *Staphylococcus* spp., *Erysipelothrix rhusiopathiae*, *Streptococcus* spp., *Clostridium* spp. and *Mycobacterium* spp.

there are those species that are from the family Enterobacteriaceae, which include *Edwardsiella trade*, *Klebsiella* spp., *Salmonella* spp., *Yersinia* spp. that are in most cases isolated from the aquatic environments given the fact that they are normally associated with human illness. Scientists claim that fish parasites and toxins are some of the things that are known to cause human illness but it is not because of the ornamental fish species, given the fact that they are not normally injected. Ideally, to ensure that there is safety for the farmer rearing the fish extensively, it is important that they ensure that a veterinary professional follow standard safety protocols and wear gloves, and also be able to advise each and every one rearing fish to take such precautions especially as they handle such fish.

In order for a breeder to be successful, there are a number of issues that he/she needs to put into consideration. As a beginning, the individual should start with working with a livebearer before deciding on an ornamental is to rear. After working with a livebearer, the goldfish is the second type of ornamental fish or any other type of egg-laying ornamental fish. This will be very helpful in ensuring that the farmer is acquainted with the investment, which involves the handling and the maintenance of the brood fish and their young ones. As such, it would be very helpful to obtain knowledge on the biology, feeding behavior and good conditions of the fish are the most important aspects in rearing ornamental fish. There are live foods for the ornamental fish which include earthworms, sludge worms and bloodworms among others are known to be stock for the brood and as such, the larval stage of the same need a lot of attention.

Figure 3.8: Live foods for the ornamental fishes.

Source: European Scientist

In order to ensure that a unit of ornamental fish is continuously provided with food, it is necessary to maintain the unit successfully. Breeding is considered easy in most cases but larval rearing is known to require special attention. Local agro produce are known to be important supplementary feedings for farmers as they can be able to prepare on-site pelleted feed. In order for the farmer to ensure that they can avoid health-related problems, they need to ensure proper water quality, which can be done by installing bio filters. Breeding and rearing units need to be made close to a regular delivery of water and electricity. If the unit is positioned close to the streams, it'll be incredible wherein the unit can get hold of running water and the rearing unit may be made flow-through (Ellingsen et al., 2009).

Constant availability of agro-byproducts like oil cakes, rice polish and wheat bran, and animal-based protein consisting of fish meal and prawn-head meal, will facilitate instruction of pelleted eating regimen for the fish. The brood-inventory decided on for breeding needs to be the best in order to ensure that they produce the best fish for sale. It is useful that the younger ones may be grown until their maturity. It is now no longer most effective for sufficient benefits in managing the fish however it additionally allows in managed selection. Breeding and rearing units can be mounted ideally closer to airport/railway stations in order to ensure that live fish may be transported to inner marketplaces and additionally for export without difficulty. A fish breeder may want to pay attention ideally on a single marketable species to streamline control measures. Proper expertise on marketplace demand, purchaser desire and overall, the operation of an advertising and marketing community run on public contacts and public relations is desirable. Pioneer and professional companies on this discipline can also be continually kept in contact with to follow the arrival of new trends in advertising and marketing in addition to technical advancements, through training.

Figure 3.9: Fish industry recognizing ornamental fish trade at the second international ornamental fish trade and technical conference.

Source: FAO

3.4 ORNAMENTAL FISH REARING

Ornamental fish are fish kept or reared for their colorful or peculiar attractiveness. They are used for decorative purposes in for example houses where there are kept in an aquarium. Recently, the ornamental fishes have brought significant change to the fish industry, as they have been increasingly involved in the international commercial trade. This has therefore led to increased earnings from the foreign exchange. The breeding of these ornamental fish began about 1000 years ago in the world. Firstly, there was the domestication of the goldfish, which was later followed by other several fish species of which some are very pricy because of their distinct markings in their bodies. For example, the koi carp fish is among the expensive ornamental fish. In India however, Ichthyo fish species such as *Botia dario*, *Botia rostrata*, *Colisa fasciata*, *Trichogaster labiosa* as well as some species of *Nemachelilus* and *Schistura* fish are considered ornamental. Besides, in the United Kingdom, several studies have reported that in every 10 households, it is estimated that one possess fish as a pet in their house. Additionally, in the United States, more than 20 million fish are kept in the aquaria in people's houses while there are about 20 million pounds in the United Kingdom. Even though there is only 3% of fresh water in the world, generally approximately 4000 freshwater as well as marine species are kept as pets at residential places or as exhibits in public spaces and rooms. Pet fish are very diverse and are reared from a variety of natural habitats. These also are the case of ornamental fish (Edwards, 2015).

Generally, like most fish species, ornamental fish are wild-caught. They take approximately about 3 days to be transported from the areas of their catchment such as in the Amazon and or the coral reefs. During transportations by ship of the wild-caught fish species the mortality rates are estimated to be ranging from 1% to 30%. These are a result of the various stresses that fish are exposed to as they experience poor conditions as well as acclimation even as some go through their life changes stages. Nonetheless, the death rates are reported to be higher after the delivery of the fish to the new environment. Globally, the ornamental fish are of significant contribution to economies and are estimated to be of about GBP £3 to £4 billion value. In the retail market, a single pet fish ranges from around one pound (GBP) for a goldfish for example to about 10000 (GBP) in the case of top-quality koi.

Most fish farmers have adopted the rearing of ornamental fish given its profitable margins. In fact, it has been reported that 90%of the freshwater ornamental fish are farmed, leaving only 10% which are collected from the wild areas. In areas such as Israel and the Far East, farmers engage in ornamental fish production on a large-scale basis. This has resulted in high levels of pricesafety concerns in the cases of diseases and mortality in these areas. There have been reported cases of viral outbreaks from the various intensive fish-rearing regions recently in the world. This is because of the large amounts of fish that are wild-caught as compared to those that are reared. Several ornamental marine fish have been produced for marine trade fish that have affected the wellbeing of humans as well as have negatively affected the conservation of biodiversity as a whole. These are the fancy bubble eye goldfish and the balloon mollies, which are characterized by features of the stunted body, shapes as well as distended abdomens. They particularly cause ill health as well as stress to humankind. There are also a number of hybridized fish species that are reared for ornamental appeal such as the parrot cichlid.

In ornamental fish rearing the activities involved are inexpensive in that any adult would engage in by providing the water and food requirements necessary to ensure production. They are unlike other animals that would require animal husbandry care as such as livestock keeping. From reputable ornamental fish stores or pet shops, potential buyers may be able to enquire on the particular requirements for ornamental fish rearing including the size of the tank to be used, water conditions as well as other factors which are relevant according to the fish species to be kept. This is however not very enterprising if the owners are not in a position to acquire the skills needed in the fish business. On the other hand if well invested by skilled personnel, is has been considered as a multibillion dollar industry which is easy and a stress relieving hobby to its buyers.

In the World, Asia is the biggest exporter of ornamental fish. Of the ornamental fish, about 90% are traded from Kolkata, 8% in Mumbai while about 2% are traded from Chennai. In India, ornamental fish are only about 0.008% of the global trade, which is estimated to be Rs.158.23 lakh. The industry has potential for growth and development as it is accompanied by high returns from short life spans and increasing demand. In Himachal Pradesh of the fifty-eight ornamental species, they belong to 5 orders, 13 families as well as 36 genera (Díaz et al., 2019). Some of the indigenous ornamental fish are almost becoming extinct due to neglection in the rivers due to a lack of awareness. The indigenous species can also be collected from their wild habitats cultured and reared so as to avoid their extension as well generate profits.

For the ornamental fish, their prices vary with factors such as consumer preferences, sex, size, species availability as well as the ease associated with shipping and culturing. For instance, the males with brightly colored bird wrasse or also known as *Gomphosus varius* may be sold twice as much as the duller or its counterpart same size female species. Alternatively, the blue devil which is basically an orange-tailed male or also referred scientifically as *Chrysiptera cyanea* may sell at twice the amount fetched for the female counterpart. On the other hand, small clown trigger fish referred to as *Balistoides conspicillum*, fetch more price as compared to the extra-large. In the fish availability factor, reality means that the fish species sell at higher prices often. For example, the angelfish are known to exceptionally sell at hundreds of dollars in the United States markets. Particularly, cultured fish acquire small markets in the conservation minded aquarists customers. The ornamental fishes may be as twice expensive as compared to food fishes. Several studies have concluded on the possibility of the ornamental fish industry depending on the wild fishes given that the economic realities or ornamental fish rearing and culturing are few as well as reduced number of species available as well as suitable for culturing.

In practice, ornamental fish rearing, it is imperative for the farmer to comprehend several attributes such as the various breeding behaviors of the fish, sexual differentiation, larval rearing, water chemistry, nutrition as well as diseases among other factors. Relatively, there is less srate required for aquaria keeping. The space that has been set for the aquarium should have diffused light. Being that they are beautiful, the enterprise has attracted many people who practice it as a hobby and for adventurers who would like to find out about fish farming. In India, which is also among the highest in producing fish exports of ornamental fish, are mainly confined to mostly the freshwater varieties, which are limited to the fish that are collected from the wild nature found in the North-eastern states. This accounts for 85% while 15% are generated from few bred exotic species. In the states breeding and rearing of ornamental fisheries is not largely practiced. There are only a few farms in the region that are reported to have engaged themselves with ornamental fish breeding although it happens on a trial basis. This has been besides the great demand for fish as people increasingly have increases on aquariums decorations at their houses as well as their places of work.

Most ornamental fish breed well when exposed to temperatures that are warm. For the case of other exceptional such as the goldfish and koi carp ornamental fish, they bred well in hilly areas where the temperatures are low especially during the winter period. The process of breeding ornamental fish requires simple equipment or instruments as well as convenient aquarium breeding sites. The site may be located in a spare tank. The only requirement that a farmer should put in place is to ensure that all the activities carried out are in line with the habits as well as the biological requirements of a fish. A hobbyist is said to have the ability of studying fish behavior and their biology in aquarium maintenance and breeding of fish varieties. Factors that are highly important in the rearing of ornamental fish include factors such as water quality .This entails the water temperatures, lighting in the waters and its density. Ornamental fish are generally 3 to 4 cm long, especially for the tropical ornamental fish varieties. They thus may fit in considerable numbers in a moderate aquarium size and breed effectively. The aquarium or ornamental fish are grouped into two categories based on their breeding behavior, egg laying and livebearers. The most important groups of egg layers are barbs, goldfish, tetra, danio, betta, cichlid along with gorami. Widely known livebearers include guppy, platy, molly and swordtail plus their variants. Among the egg layers in the ornamental fish are the Barbs. They are reported to have originated from India including varieties such as rosy barb (*Pythia conchonius*), tiger barb (*Puntigrus tetrazona*) striped barb (*P. fasiatus*), Aruli barb (*P.arulius*) and Narayan barb (*Pethi narayani*). Other examples of ornamental fish generated are Danios (*Danio aequipinnatus*), pearl danio (*Danio albolineaus*) and Zebra danio (*D. rerio*). The most commonly easy type of ornamental fish in terms of rearing and breeding is zebra danio of Indian origin (Diana et al., 2013).

The Goldfish which is scientifically referred to as *Carassius auratus* is the most common pretty fish, preferred by most hobbyists for the reason that its eye-catching with coloration ranging from pure gold to red, orange black and albino to a range of shapes as well as different sizes. common varieties of goldfish available include are comet goldfish, lion head, oranda, fringe tail, veil tail, fantail, shubunkin, telescopic eye, Lichi gold etc. Gold fish grow up to 20 cm in length, except they start breeding when they are 6 cm in most cases. Goldfish tetra is known to be small fishes ranging from 3 to 8 cm in length. The majority of the tetra are reported by past studies to have South America as their origin. Common species of the tetra include black widow tetra (*Gymnocorymbus temetizi*), Serpae tetra (*Hyphessbrycon callistus*), rosy tetra (*H. rosaceus*), lemon tetra (*H. pulchripinnis*), flame tetra (*H. flammeus*), neon tetra (*H. innesi*), cardinal tetra (*Cheirodon axelrodi*), glowlight tetra (*Hemigrammus erythrozonus*), head and tail light tetra (*H. ocellifer*) and pretty tetra (*H. pulcher*). *Betta splendens* universally known as Siamese fighting fish or betta is an additional type of ornamental fish species that occur in varied colors like green, red blue, albino and occasionally with a mixture of extra shades. The male ornamental fish are found to be brightly colored as compared to the female ornamental fish. For this reason, the male ornamental fish are most preferred. The male ornamental fish have beautifully spread over fins to attract most hobbyists. During their growth, they are aggressive in the presence of other male fish.

Moreover, Angelfish is an important ornamental species usually favored of diverse varieties for example black, veil tail, marble and albino. Livebearers are other of ornamental fishes which are ascribed to giving birth to young ones and reproducing only a few numbers of offspring in comparison to egg layers given that their breeding is rather effortless. In the livebearers, ornamental fish the growth and development of the young offspring happen inside the female fish and they are released after an average of four weeks or in one month's time. Some of the species in livebearers include guppy (*Poecilla reticulata*), black molly (*Poecilla sphenops*), swordtail (*Xiphophorus helleri*) and platy (*X. maculate*).they can produce up to about 60 offspring at a given time period. The livebearers with larger swordtails may reproduce up to 100 offspring. This is the case also if the female fish is fed properly with natural foods, which should be supplemented, with quality artificial feeds. Farmers are emphasized on the importance of separating the young ones from their mother as soon as they are released so as to avoid predation or cannibalism of the young ones by the mother (Diana, 2009).

The common culturing facilities for ornamental fish production include cement cisterns, glass aquaria, earthen ponds as well as earthen pots. About 3 to 4 cement cisterns are enough for small-scale ornamental fish productions. The cement cisterns may have dimensions measures of 3 x 2 x 1m, which should be built, slightly above the ground for easy drainage. For breeding purposes by the use of aquaria, glass aquaria are preferred as they promote easy usage of heaters and aerators. In other instances, fish farmers use small earthen tanks as rearing juveniles while incorporating the food fish. Marginal farmers have been found to use large earthen posts with average approximate diameters of 1.5m in culturing larvae as well as juveniles of ornamental fish. Scientists have concluded that rainwater is in fact the most preferred source of water with ornamental fish rearing. This is because unlike rainwater most municipal water is chlorinated thus not suitable for ornamental fish production. It is expected therefore that if the municipal water is to be used then the water should be kept in an aerated space for several days so that it can de-chlorinate before usage. Tube well water in the rural areas has come in handy in ornamental fish rearing in remote regions. The average temperatures that are required in ornamental fish rearing is between 15 and 28 °C. The pH of the water should be slightly alkaline.

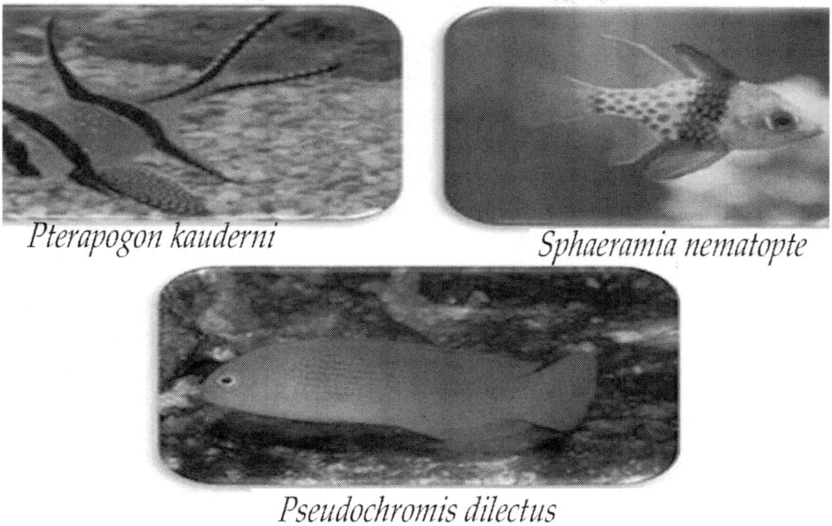

Figure 3.10: Breeding and rearing of marine ornamentals.

Source: ResearchGate

Generally, different ornamental fish require different specifications of factors imperative to their breeding, growth as well as their development. However, most of the species survive well on medium hard waters. Additionally, a shade facility is of great importance above the facility or the equipment used to enhance breeding and rearing of ornamental fish. Sufficient water and electricity supply are additional considerations that are requiring for successful ornamental fish rearing at their locations. The farmer is also expected to select a quality brooder that will lead to enhanced fish breeding activities. Livebearer ornamental fish prefer to breed in water that is hard and additionally alkaline. On the other hand, the Angel, Discus, Tetra, Oscar and Loaches ornamental fish prefer soft acidic water while goldfish, danio, catfish, rosy barb, and fighter fish tolerate a wide range of water qualities. So as to ensure continuous increase in the breeding activities the farmers should breed a fish for at least two years at most. This will also be advantageous in that it will promote increased breeding viability. The farmers may achieve this by collecting their fish from different places in the ecosystem.

Investors in ornamental fish rearing activities could be able to make farm made aqua pellets by the use of different agricultural by-products especially by using small pelletizes. The farm fed may also be manufactured by the use of commercial fish or prawn feed available in the markets. Other foods that may be used to the ornamental fish in promoting brood and larvae include live food like Zooplankton, Tubifex, Bloodworms and Chopped Earthworms. The small-scale farmers might not afford the different made packed pellet feed for the ornamental fish because of their associated high costs. Therefore, the prospective farmers are mainly encouraged to get training and new demonstration methods entailing feed preparation procedures using locally available materials. At the farm level, good marketing of fish is required as well as proper communication facilities with small, middle or large traders at the farm level. Some of the lowly cost alternative ornamental fish feed include water fleas, Tubifex or sludge worm, mosquito larvae and chopped earthworm. Homemade feed for the fish may be as whole-wheat bread, vegetable peelings as well as rice. Commonly, most ornamental fish farms have been found to depend on Daphnia, Tubifex worms and mosquito larvae for feeding the fish. Tubifex worms and mosquito larvae for feeding the fish are basically collected from the sewage water channels in the neighboring ecosystems and used as food for the ornamental fish. It is expected that the farmer should feed the fish at least once every day. Ideally overfeeding of the ornamental fish is considered more life-threatening as compared to underfeeding. This is because overfeeding in fish results in contamination of water thus the water quality may not be favorable for fish rearing. In the industry, fish food bears almost all the operational costs associated with culturing or rearing ornamental fish. The use of live food as feed is advocated as it is said to contain all essential nutrients required by a fish in ensuring higher breeding efficiency levels and growth to vibrant colors. In conclusion, in ornamental fish rearing, the most important physical and chemical qualities of water that could influence aquatic productivity are such as temperature, pH, dissolved oxygen, free carbon dioxide, total alkalinity, water hardness, free chlorine, and ammonium. The water should further contain dissolved nutrients for example nitrogen, phosphorus, potassium, calcium, magnesium and silica (Diana, 2012).

3.5 ORNAMENTAL SPECIES OF FISH

The Guppy, which is also known as, million fish or in other areas the rainbow fish and it is one of the world's most distributed tropical fish. It is also among the most wildly distributed freshwater aquarium fish. The scientific name is *Poecilia reticulata*. It is a member of the family Poecilidae and according to research, the family is live bearing. They are known to have originated from the Northeast of South America but they have now been introduced to many habitats and are currently found across the world. They are found almost everywhere because they are highly adaptive to any environment and different ecological conditions. The male guppies are much smaller than the females. They are known to have ornamental caudal and dorsal fins. The guppies that are found in the wild feed on a variety of food; these include the aquatic insect larvae and the benthic algae. The guppy is a model organism, which means it is used in the field of behavioral study, which is to understand their behavior in different environments, habitats and different stresses, evolution and ecology i.e. how they relate to the environment around them.

Guppies are native to mostly the Caribbean area i.e. Trinidad and Tobago, Venezuela, Barbados, Antigua and Barbuda, Suriname and Guyana. They are known to be found in almost all continents except in the Antarctic. They have been expected to feed mostly on mosquito larvae, which in turn will reduce the spread of malaria, but they also have an effect on the native population of where they are introduced. The guppies are known to exhibit sexual dimorphism that means the two sexes of guppies exhibit differences in their characteristics apart from the difference in their sexual organs. In the wild, the females are grey in body color while the males have splashes, stripes or spots and can be from a wide variety of colors.

Figure 3.11: The Guppy. They are of different colors and they have different sizes of fins. The tailfins sometimes appear like snakeskin.

Guppies are also produced by breeders through a process referred to as selective breeding, which is characterized, by different colors, shapes, patterns and the size of the fins which can include snakeskin and some that appear like grass. The mating system in guppies is polyandry where the female is able to mate with several males. The males are sometimes seen to only provide mating services as they do not provide any material benefit to the female and they also do not provide parental care to the young.

Guppies prefer to be kept in aquariums with hard water and the temperature should range between 25.5 to 27.8°C and the salt level of the water should be that equal to one tablespoonful per every 19 Liters of water. They are generally peaceful but sometimes they can be seen to nip i.e. between male guppies and sometimes against fish that have large fins such as the angelfish. The males and females should both be kept in the aquarium as both show shoaling. When the adults are well fed, they do not feed on their young but the fry should be secluded from the adults (Cressey, 2009).

3.5.2 Siamese Fighting Fish

It is commonly known as the betta and it is a freshwater fish and is native to Southeast Asia. It is native to Laos, Myanmar, Indonesia, Cambodia, Vietnam and Thailand. They are scientifically known as *Betta spiendens*. Out of the 73 known species of the betta fish from the genus Betta there is only one i.e. *Betta splendens* which is eponymously known as "betta" and this is because they are globally popular and are known to be pets and are among the most widely available aquarium fish.

Figure 3.12: Betta or the Siamese fighting fish. They are brightly colored and have large fins.

In addition, this is because they have varied and vibrant colors and they are also known to have diverse morphology and they have relatively low maintenance. They are known to have been domesticated for the last 1000 years, which is among the longest of any fish. It is the national aquatic animal of Thailand and the country remains as the primary breeder and exporter of the Siamese fighting fish. Although they are kept in abundance as pets, they are classified as vulnerable because of pollution and destruction of their natural habitat.

They came to be known outside of Thailand courtesy of King Rama III who lived between 1788 to 1851. It is said that the King gave some of the fish to Theodore Cantor who was a Danish Physician, Botanist and Zoologist. They are known to first appear in the west in the 19th century and within a few decades, they became popular as ornamental fish.

These fish are known to be territorial and males are known to attack each other if they are kept in the same tank. Females on the other hand can also become aggressive towards each other if kept in a confined space. If the fish do not get a place to escape then it can lead to the death of one or both of the fish. These fish are known to be tolerant in waters that contain low oxygen and poor waters and this is because of their labyrinth organ, which allows for the intake of surface air.

They usually grow to the length of about 6 to 8 cm. the ones that are kept in aquariums are known to have brilliant colors and large flowing fins whereas the ones found in nature are green, brown and grey and their fins are generally shorter. These fish prefer water of temperatures between 24 to 28°C but have been seen to survive in waters of about 13°C. However, in most instances, old water weakens their immune system and this will make them susceptible to certain diseases. Water should also be of neutral pH (Chavanne et al., 2016).

It is recommended that individuals need to monitor how the bettas interact with other fish, the fish in the tank should be tropical fish, no territorial, communal, and should not have a similar body type to the bettas or the fins should not be flowing. Species such as tetras and danios are preferred because they keep to themselves and can endure the territorials of the betta with their larger number. These fish are quite intelligent and can be quite inquisitive and hence they will require stimulation and without it, they end up being bored and depressed and this will make them have a weaker immune system. The aquarium should contain decorations such as plastics, rocks, caves, driftwood and other ornaments.

3.5.3 The Goldfish

This is a freshwater fish that is scientifically known as *Carassius auratus* and is from the family Cyprinidae and the order Cypriniformes. It is mainly kept indoors as a pet in indoor aquariums. It is one of the most popular aquarium fish. In some parts of North America, the goldfish that have been released into the wild have become invasive and can even grow to the size of a full-grown tilapia. They are native to the east of Asia. They are a small member of the carp family. It was first bred for color in imperial china approximately 1000 years ago. Since that time, there have been several distinct breeds that have been developed. These goldfish come in various shapes, sizes, fin configurations and come in different colors i.e. yellow, white, black, orange, red and brown.

When the goldfish are kept in an aquarium, they can only grow to about 2.5 cm to 5.1 cm. They have been known to grow larger when left to roam free and a known example was a goldfish found in Holland and was 48 cm in length. They are one of the fishes that are being studied for their vision because they are known to have four cone cells, which are sensitive to different colors i.e. green, blue, red and ultraviolet. They are classified as tetrachromats because they can distinguish between four different primary colors. They are also the most studied fish for their hearing senses as they have two otoliths that allow them to detect the sound particle motion and the Weberian ossicles which connect the swim bladder to the otoliths and this allows them to detect sound pressure.

Figure 3.13: Goldfish. They can distinguish four different colors and are intelligent as they can know individuals and have memory for up to three months. They can also be taught tricks.

Source: Fish Care

When provided with the right nutrition and enough water they will grow to sexual maturity. They usually breed when there is a significant temperature change and this is often in the spring. The males chase the females carrying the eggs i.e. the gravid fish, bump, and nudge them to prompt them to release the eggs. Within a week, the eggs hatch and they can take up to a year for the color to fully become the goldfish color, before that they are metallic brown, which is similar to their ancestors.

They are considered friendly fish as they display schooling behavior. They are known to have varied breeding; feeding and they also avoid predation, which has largely contributed to their success. They can only compete for food when it becomes scarce. They have associated learning skills and they have social learning skills. Their visual acuity allows them to distinguish between individual humans (Cole et al., 2009). The owners notice that they will come close to the front of the glass or they can swim rapidly around the tank and sometimes they go to the surface and mouth to show they want food which generally shows the person is favored by them but sometimes they hide when others come to the tank. They have a memory span of about three months and they are able to distinguish between shapes sizes and different colors. When they are positively reinforced, they can be trained to recognize and react to different colors and signals, which simply said they could be trained to perform tricks.

3.5.4: Angelfish

It is known by aquarists as angelfish but they are known as Perophyllum, and they are a small genus of fish that are found in freshwater and is from the family Cichlidae. They are native to South America in the Amazon basin, Orinoco basin and some rivers in the Guiana shield. They are unusually shaped as they are laterally compressed with round bodies and they have a triangular anal and dorsal fin. Their body shape is useful as they are able to hide within plants and roots. The angelfish that are found in the wild are stripped transversely and some coloration that allows them to camouflage to the surrounding surface. They are ambush predators and mostly feed on small fish. They form monogamous pairs. Their eggs are laid on submerged logs or a flattened leaf.

Figure 3.14: Angelfish. They are unusually shaped as they are laterally compressed with round bodies and a triangular anal and dorsal fin.

Source: BeChewy

They are also one of the most kept fish in aquariums and as well as the most kept cichlid. They are praised because they have a unique shape, color and behavior. It was in the late 1920s and early 1930s when it was first bred in captivity in the United States. The most commonly kept angelfish in aquariums is *Pterophyllum scalare* and they are mostly captive-bred. They are kept in warm aquariums preferably 27°C. They are generally known to be peaceful when they are not mating but, the big fish will always feed on the small fish.

They reach sexual maturity between 6 to 12 months depending on the conditions. In case the eggs are removed from the aquarium, the pair can produce viable eggs every 7 to 10 days. At around three years, the spawning decreases and then eventually stops. When ready to spawn the monogamous pair will choose a place for them to lay the eggs then spend one or two days picking out algae and detritus from the surface. The female then lays the eggs on the spawning area then the male fertilizes the eggs, they repeat this process until they have a total of between 100 to 1200 eggs and this will depend on the size and age of the female. Both of the parents will care for their young during their development. They maintain a high rate of water circulation around the eggs by swimming close to the eggs and fan them using their pectoral fins (Costa-Pierce et al., 2021).

They are placed in the semi-aggressive category in the pet store. They can be kept with other fish such as tetras and barbs but those that will be small enough to fit into the angelfish mouth will be eaten. There should be enough food in the tank so that the angelfish will not end up feeding on their tank mates.

3.5.5: Koi

In Japanese, it is known as nishikigoi and they are the colored varieties of the Amur carp. They are mostly kept for decorative purposes i.e. in koi ponds or in water gardens in the outdoors. They are scientifically known as *Cyprinus rubrofuscus*. There are many varieties of ornamental koi. Their breeding began in Niigata, Japan in the 19th century. They are distinguished by coloration, scalation and patterning that enables the Japanese to recognize the different varieties. Some of the common colors of the koi include; blue, orange, black, cream, yellow and red.

Figure 3.15: Koi are mostly kept for decorative purposes outdoors in water gardens or ponds. They are also some of the most expensive aquarium fish in the world.

Source: www.nippon.com

These types of fish are native to central Europe and Asia. They were especially domesticated in the east of Asia where they were mostly used as food. The carp are mostly Coldwater fish, they can adapt too many climates and different water conditions, and this allowed the species to be propagated to many different locations including japan. The fish were first bred for color mutations in china and that is almost 1000 years ago. The systematic breeding of the ornamental koi in Japan began in the 1820s in Ojiya and Yakamoshi in the northeast coast of Honshu. The breeding first gave rise to the red carp, which was then followed by the pale blue asagi and white, then the red and yellow bekkou. The next was the sarasa variety, which had a red and white pattern and was created in 1830. The single-colored metallic koi is the most recently developed and is regarded as the ogon variety.

The keeping of koi as a hobby has spread worldwide and has become a social hobby. It has also led to the formation of koi clubs where hobbyists who are passionate share their knowledge and they help each other with their koi. In the 21st century, some of the wealthy Chinese imported koi from Niigata, Japan and the price of the carp is quite high. In 2018 a Chinese collector bought one carp at approximately $2 million dollars, which is the highest price ever recorded (Amenyogbe et al., 2018).

There are new koi varieties that are still being developed. In the United Kingdom the ghost koi have become quite popular and it was developed in the 1980s. There is the butterfly koi, which is also known as the longfin koi or the dragon carp and was developed in the 1980s and they can be notices by their long and flowing fins.

The keeping of koi in outdoor water gardens has become quite popular among many affluent Chinese. In Japan the koi is seen as a symbol of prosperity, good luck and good fortune and in some communities, they also see it as a symbol of perseverance in times of adversity. It has also become popular in other countries such as Sri Lanka where they are kept in interior courtyards.

3.5.6 Zebrafish

They are freshwater fish that are found in the minnow family i.e. Cyprinidae and the order Cypriniformes. They are scientifically known as *Danio rerio*. They are native to South Asia and are a popular aquarium fish and sold under the trade name zebra danio. It is very important and scientists use it also as a vertebrate model organism while doing their research. They are mostly used in drug development especially pre-clinical development.

Figure 3.16: Zebrafish. They are regarded as zebras because of their five uniform horizontal and pigmented blue stripes on their back, which extend to their caudal fin.

Its habitat in South Asia is quite large i.e. India, Nepal, Bangladesh, Pakistan and Bhutan. It has recently been introduced in California, New Mexico, Florida and Connecticut in the united states and this has been done by aquarists or some have been released by fish farms. They are mostly found in moderately flowing or in stagnant water that is clear and quite shallow and this can be in ponds, canals, streams, ditches oxbow lakes, rice paddies and ponds.

They are referred to as zebrafish because of the five uniform, horizontal, pigmented blue stripes on the sides of their body and they extend to the caudal fin. Her shape is fusiform and are laterally compressed and their mouth is directed upwards. The males are torpedo-shaped and they have yellow stripes in between the blue stripes while the females have a larger whitish belly and silver stripes instead of gold. They can grow to up to 4-5 cm in length while in the wild and it will also depend on the location. Its lifespan while captive is 2-3 years but in ideal conditions, they can live to up to five years (Colsoul et al., 2021). For a female to ovulate and spawn a male must be present. The female s able to spawn at an interval of two to three days and they lay a clutch of 100 eggs. The embryo begins to develop even in the absence of sperms but it stops after the first few cell divisions. The fertilized eggs almost become transparent immediately.

They are considered as hardy fish and are good for beginner aquarists. They are rapid breeders, are aesthetic, cheap and are broadly available. They do well in schools or even in shoals of six or even more and they also interact well with other species of fish in the aquarium. They are usually susceptible to velvet disease, microsporidia and mycobacterium. The adults, when given an opportunity, can feed on their young i.e. the hatchlings and this can be controlled by separating them by using a net, breeding box or they can be kept in separate tanks. In captivity, they can live for forty-two months i.e. three and a half years. They have been used to make genetically modified fish and the first ones were sold as the GloFish or the fluorescent colored fish.

3.5.7 Gourami

This is a group of freshwater anabantiform fish that are of the family Osphronemidae. They are also referred to as Gouramis or Gouramies. They are native to Asia and they stretch from Pakistan to India, Southeast Asia up to the northeast towards Korea. The name Gourami is of Indonesian origin and is sometimes used for the family Helostomatidae and Anabantidae. Many of them have an elongated, feeler-like ray at the front of each pelvic fin. They all show parental care to their young. Some of them are mouthbrooders i.e. oral incubation which is they give care to their young by holding them in the mouth of the parent for an extended period. Almost 133 species have been identified.

In the areas of Malaysia, Indonesia, Brunei, Thailand and the Philippines they fish the gouramis from streams, rivers, canals and brooks and sometimes in larger water systems. Most of the gouramis species are aquarium fish and they are widely kept in the world. They are labyrinth fish and hence they will swim near the top of the tank. They can feed on prepared food or live feed. There are some species that can become quite large and hence cannot be kept in small aquariums. As labyrinth fish, they have a lung-like labyrinth organ that allows them to gulp for air and use atmospheric oxygen. This organ is a very vital organ for fish that are often found in warm, shallow and oxygen-deprived water.

Figure 3.17: Gourami. They are generally peaceful and are known to be mouth brooders i.e. they take care of their young in their mouths.

Source: Discovering all marine species.

Gourami are generally regarded to be peaceful but they are also capable of harassing and even killing small or long-finned fish. The gouramis males have been known to spar with each other and if there is overcrowding there can also be aggression in the tank. The compatibility in the tank will be determined by the species of gouramis and the fish it is housed with.

3.5.8 Southern platyfish

They are also known as common platy or the moonfish and they are a species of freshwater fish from the family, Poeciliidae and the order Cyprinontiformes. They are scientifically known as *Xiphophorus maculatus*. They are livebearers, which means they can retain their eggs inside their bodies and then give birth to live, and free-swimming young. They are native to north and Central America and stretch from Veracruz in Mexico to the north in Belize.

They grow to a maximum length of six cm. they exhibit slight sexual dimorphism and the male caudal fin is more pointed. The male's anal fin has evolved into a monopodium, which is a stick-shaped organ that is used for reproduction. The female's anal fin is fan-shaped. The wild southern platyfish is drab in its coloration. Breeders have been developing them with various color varieties such as red/black, black/white, red, yellow and orange which are mainly kept in aquariums and by hobbyists.

Figure 3.18: Southern platyfish. They are a hybrid of two different species from the same genus.

Source: Aquarium Tidings.

They are easy to keep and they can live as a community with other fish. The water's pH should be between 7.0 to 8.0 and the hardness of the water can range between 9.0 to 19.0 dGH whereas the temperature will range between 18 to 25°C (Bjelland et al., 2015). When they are kept in captivity, they mature in about three to four months and then they readily breed. The females give birth to about 20-40 young at one given time. Adults and other fish can eat the hatchlings but in case, there is gravel then the young can hide and survive because they are known to be hardy fish. The fish that are sold in pet stores are not the pure strain of *X. maculatus* but they are a hybrid of *X. helerii* and *X. maculatus*. The males have a sword-shaped tail and hence they are called the swordtails. The color and fin shape will vary wildly in the aquarium trade as it all depends on what the consumer or hobbyist will prefer.

3.5.9 Ticto barb

It is a species of subtropical freshwater fish that belongs to the family Cyprinidae. They are also known as two-spot barb and their scientific name is *Pethia ticto*. They are native to the upper parts of the Mekong river, Mekong, upper Charo Phraya, Salween and Irrawaddy. These areas are found in Nepal, Pakistan, Myanmar, Thailand, Bangladesh, Sri Lanka and India. It is sometimes confused with the Odessa barb in the aquarium trade but the difference is in the male of the Odessa, which is reddish-orange, which is not a color in *P. ticto*. They are silver, gold and they have two spots, whereby one is found just before the pectoral fin, and one spot is near the back of the tail. It grows to about 10 centimeters in length. They are usually found in waters that are still, shallow, and marginal waters of lakes and rivers and these areas usually have a muddy bottom. They prefer water with ph. between 6.0 to 7.0 and the water hardness should be 10 digs. The temperature of the water should range between 14 to 22°C. They majorly have a varied diet ranging from plankton, insects and small crustaceans (Bentzon-Tilia et al., 2016).The ticto barb is an active schooling fish, which means they are usually kept in groups. When they are in a large enough group, they wot bother any other species of fish. They are egg layers and prefer to lay them on a coarse gravel bed. When spawning they lay quite a large number of eggs i.e. 150 eggs and they will lay 20 eggs at a time. As soon as spawning is finished, they feed on their eggs which means they need to be separated from the eggs as soon as they finish spawning.

Figure 3.19: Ticto barb. They are usually silver and gold and they have two spots on their body.

Source: Tropical fish keeping.

Chapter 4

Fish Farms Within Extensive Aquaculture

CONTENTS

4.1 INTRODUCTION

This usually refers to fish farming that is conducted in medium to large-sized ponds or can be done in water bodies. The production of fish in such systems depends merely on the natural productivity of the water. However, they can be slightly or moderately enhanced. Inputs that are supplied from external sources are limited and for this reason, costs a normally kept low, there is a restriction in capital and there are cases where the quantity of fish that is produced tends to be low. These fish are not provided with supplements and as such, the return on labor is quite high. This is unlike intensive fish farming where the quantity produced per unit area is quite high (Bottema et al., 2021). In order to be able to intensify the culture, the various production factors such as the water quality, the feeds provided to the fish and the quality of the fingerlings. These aspects need to be controlled in order to improve the production conditions. For this case of fish farming, there needs to be steady monitoring during the production cycle.

Figure 4.1: Extensive fish farming. The stock feeds on natural products such as algae, zooplankton and crustaceans.

It is with no doubt that, these aspects involving the rearing of fish in an extensive manner require that the farmer use high technical practices and have high capital investments. In the process, it will add to the production costs. However, the returns that are accumulated need to justify the increased production costs. Natural productivity contribution to fish production is normally either very low or negligible. It goes without doubt that intensive fish farming is connected with high costs or has threats to the environment. the extensive form of fish farming and intensive are the most common ones though there are others which include semi-intensive fish farming which has high costs or have a couple of threats to the environment. This method takes elements of both intensive and extensive fish farming though defining the same in this manner is described to be under definition of the same (Verdegem

& Bosma, 2009). Before deciding on which form of the above-stated types of fish farming to use, the farmer needs to be very conversant with all the methods and knows the costs they will incur because of the same. Another aspect involves looking at the context under which fish farming is practiced. The most important variable to be considered is whether the production system is ensuring food security for his/her household. It is important that they secure this aspect at the first point.

As an individual looking to ensure that food security for his family is sustained, growing of food crops should be seen as an important aspect and then followed by diversifying their activities through venturing in other activities ranging from the agricultural to the non-agricultural aspects. Venturing into such is because of trying to pursue various income-generating activities where the rationale behind the same is the management of risks before engaging in the business and how to cope with losses once they have started the activities. The strategies that the farmer decides to devise needs to take into account the assets present at the place the entrepreneur is conducting their business. Furthermore, the farmer needs to look at the investment capacities. For the case of intensive fish farming, the investment required for the same is fairly higher compared to the investment required for the case of extensive farming of fish. To be able to achieve all the stated objectives, the farmer needs to also consider the marketing opportunity for his fish. Take for instance the case of food or labor scarcities; one would see that the performance of their business has been altered. As such, the farmer will be prevented from achieving their objectives. In order to support the various strategies that the farmer has, essentially, the farmer needs to optimize the use of resources that are within his reach. As such, the main objective of going into fish farming is the expected returns the farmer has for his business.

Figure 4.2: A fish market is one aspect that needs to be considered when choosing a system of fish farming.

Source: WGBH

In the case of fish farming, the intervention of the farmer is normally concentrated n the reproduction of stock. Extensive practices are known to be old and date back to the 13th century. In Europe, the fattening of mussels on stakes dates back to the 14th century and has been a continuous process since then. The price of shooting spat as a treatment for the overexploitation of naturally produced oyster beds became realized as early as the second part of the nineteenth century. The price of transplanting younger ones from their nurseries within the south of the North Sea to much less densely populated lagoon or sea fattening zones, at the start of this century. The mastery of developed enthusiasm for sea farming: the primary releases of younger salmon from hatcheries date from the middle of the remaining part of the century which included the parts of Europe, Pacific coasts of America and Japan. The primary cod hatchery was constructed in Norway in 1884, and the primary lobster hatchery in 1891 at the Atlantic coast of Canada (Van Doan et al., 2020).

Knowledge regarding the real manufacturing of numerous enormous structures continues to be limited. While the manufacturing of sedentary species such as shellfish and algae, whose shares are geographically wonderful from the herbal beds and normally owned by means of a charter or a company body, is reasonably properly known as the way of life for the agile species especially for the case of fish and crustaceans whose shares are exploited together. At the same time as wild shares and whose captures are not prominent from the ones of the natural populations, it is not always the case in any respect properly known. This insecurity has led the FAO to exclude from its aquacultural records the manufacturing of the same exploited shares. This selection ends in the inclusion within the fisheries records of manufacturing because of the discharge of younger fish into watercourses, lakes, coastal waters, and the ocean, despite the fact that this manufacturing could be very tedious for numerous species, especially salmon.

There is a potential for the expansion of extensive aquaculture as seen from the ecological point of view. There have been several theories that involve the reproduction strategies that involve aquatic populations, the productions levels that are achieved in certain extensive systems and the amplitude of natural fluctuations of the wild populations of fish. All these aspects considered for the case of the extensive type of farming show that the aquatic ecosystem has the capability of supporting stocks that are larger than the average biomass of the wild stock of fish. This method has been able to stimulate research given the fact that the marine ecosystem has been affected to a very small extent. In 1984, the number of marine species that were studied then was 150, which then rose to 250 in the year 1994. Researchers have however claimed that the results that were provided were disappointing. This is provided by the fact that there have been very minimal research and development programs that have led to any significant production.

Technical intensification makes it viable to exceed the productiveness limits of the wild shares. The technique includes extending the technical controls of physiological capabilities improvement of reproduction, specific stage of the lifestyles cycle of the fish, which include eggs and larvae, juveniles, adults, reproducers as well as their cultivation environments. This domestication takes part through discrete jumps: among fishing and absolutely being able to manage cultivation since there's quite a number of structures that may be labeled in four groups. The first ones are the ones, which include the reproduction process. The second ones are the ones, which additionally affect the fertility of the growth surrounding. The fourth involves the ones wherein the shoal of fish is likewise fed. The last involves the ones wherein the wastes also are recycled.

Not all the technically practicable systems are viable, and efficient, extensive methods are not always the most efficient. While technical intensification reduces the function of the natural resource, it no longer does away with the dependence of manufacturing vis-à-vis the environment. It adjusts only the character of the constraining elements and the negative impacts they impose on manufacturing. In fishing, it is considered as the tertiary manufacturing - that of the halieutic populations - which is considered as the restricting factor.

In aquacultural systems, the fish that are artificially reproduced, are currently seen to be no longer restricting. It now no longer constitutes a natural resource. Appropriated, it turns the assimilable to capital. In good-sized cultivation, it is the number one and secondary productions, which include the plankton and benthos that serve as meals for the fish, which restrict manufacturing. In extensive systems where the meals are delivered it is far considered as wastes that are unconsumed meals, faeces, antibiotics and medicines, among others, and also escapes of those that can additionally adjust the genetic historical past of the wild populations, adjusted to the capability of the aquatic ecosystem. Throughout the domestication process, the capital and labor depth of the operation grow at similar times because of the technical intensification (Turcios & Papenbrock, 2014). The growth in productiveness is considered to be in line with the unit received and accompanied through a growth within the strain in the environment. In the course of this technique, the monetary and social organization of the operation additionally is adjusted. As the technical controls diversify, policies are outlined and assist their exercise - and especially the regimes governing the possession of the natural resource and the changes occurring on the resource.

Given the fact that the fish stock feeds on the natural environment, researchers have been able to claim that extensive fish farming is more of production unlike the case of intensive fish farming that qualifies to be described as processing because of the fact that artificial food is normally provided. In the sea, outside the lagoons where fish is claimed to be cultivated semi-intensively does not exist. Those extensive systems in which the farmer aims at fattening the fish normally result in overcrowding. This case is normally similar to the biomass that is normally observed under natural conditions. The increase in biomass, which brings about the increase in total production of the stock, is normally limited by the tropic capacity of the ecosystem that is used as the cultivation environment that ideally results in the decline in yields, further lowering the production capacities of the fish stocks.

4.2 EXTENSIVE FISH FARMING SYSTEMS

The systems are based on a semi-natural environment such as in already existing ponds. There is a natural food supply for the fish and they are not provided with supplements. The fish that are cultured in such systems normally feed on crustaceans, zooplankton, algae among others. Fertilizing the pond is one way of ensuring that the pond is not out of food. One can even use animal waste to fertilize the ponds. In order to maintain the natural balance, the pond is normally stocked with a variety of fish that are from different ecological gaps. Taking for instance tilapia that feeds on algae, the catfish that feeds on the pond bed and a number of species of carp that normally feed on the zooplanktons (Troell et al., 2014). The farmer needs to be very careful with this system so as not to overstock the ponds as it may led to the reduction of productivity because of the increase in the stress levels that the fish face. This may bring about algal blooms and oxygen depletion in the ponds. Examples of systems that are used in the extensive system include the cage system, irrigation ditch or pond systems, integrated recycling systems and the classic fry farming that are discussed below.

4.3 CAGES SYSTEMS AND IRRIGATION DITCH OR POND SYSTEMS IN FISH FARMING

Fish farming enterprises rely on the design and construction of the farming structures for successful operations. Freshwater fishponds are water ponds that contain a structure of large water area and depth. The usage of the fishponds may be seasonal or perennial. Generally, the water in the ponds may be rain fed and or filled by pumping water to the pond from a nearby source. The system may require having an inlet as well as an outlet system. Water sources to a pond may be in the form of a stream or a canal as well as pumped water from underground such as water generated from wells tube wells and lakes. The levels or the ability of any pond to retain water independent of the soil composition of the pond bottom as well as the subsoil water levels where the pond is located. Besides, the productivity of the enterprise is dependent on the water and soil qualities. In small-scale fish farming the ponds used are mainly small and shallow in size. For successful production of fish, it is imperative that ponds have an assured water supply system that should have an outlet as well as an inlet. The use of ponds has been preferred as compared to other fishing systems in the fish production industry in that the method involved in fish farming is reliable and economical in use for they reduce the levels of water required in the production. Given that demand for water has been increasing tremendously from the various alternative uses, the use of pounds in fishing has been of great importance. Ponds used in fish production may serve in other production activities such as for irrigation, fish production, crop production, fire protection, energy conservation, wildlife habitat, recreation, erosion control, landscape improvement as well as for accessing water to livestock.

Largely the methods used in farming systems are diverse including extensive or intensive systems. The intensive systems may be closed or open systems in nature. Additionally, fish farming may be categorized in relation to the number of species that are kept. For example, polyculture is the cultivation of several species in a given fish system while monoculture is the production of a single type of species (Teletchea & Fontaine, 2014). Moreover, the production methods may be by the use of conventional methods where fertilizers are generally used or by organic methods, which are mainly based on the traditional techniques in production, alternatively the methods used in fish production may be integrated where the convectional and natural methods are combined. On the other hand, the use of extensive systems in fish production is mainly based on semi-natural man-made environments such as the ponds and cage irrigation ditch systems. They may contain natural food supply to the fish including algae, crustaceans and zooplankton. The cages and ponds may be fertilized using animal wastes from time to time. This is essential as it supplies the fish with nutrients that are needed for their growth and development in the production system. It is necessary that the farms with the cage irrigation ditch systems and or pond system ensure the natural balance within the method used by stocking the areas with a variety of fish occupying different ecological niches for enhanced production. For instance, the tilapia fish feeds on algae while catfish feeds off the pond bed. Moreover, some of the carp species feed on zooplankton.

Figure 4.3: Fish farms | Fish Farming information and resources, use of cages.

Source: Farms.com

In cages and or ponds, maximum care is needed in ensuring that the fish are not over stocked because of over fertilization. Lack of continuous maintenance in the fish stocks may result in fish overcrowding in the pond or cage. The situation may result in algal blooms as well as depletion of oxygen levels in the ponds, which may cause large-scale fish mortality rates. Examples of fish that are reared in a pond and or in a cage irrigation ditch system include catfish, carp and tilapia. Usually, the waters in the pond or cage should be kept at levels that ensure optimal fish production growth. In most cases, the considerable pond water depth levels are up to one meter .this is mostly the case in culturing the tilapia, carps along with shrimps. However, for the traditional milkfish ponds, the levels can be adjusted by about 40 to 60 cm of water. The fish production optimality in a pond or cage system is also dependent on the quality of water used. In the semi-intensive and intensive fish culture systems large amounts of metabolites are excreted continuously into the ponds or cages. The excess feeds that are in most cases not consumed by the fish may fill the bottom load and serve to pollute the water in the pond. The polluted water may lead to diseases to the fish because of the high modification levels resulting in mortalities. It is therefore imperative to continuously ensure freshening of water systems by allowing entry of water systems from water sources such as the river and other supply canals. The old water should be outlasted or drained by the outlet drainage systems of the sea and rivers. This practice helps to prevent the deterioration of the fishing environment within the pong or cage irrigation ditch systems.

In a high-density fish culture system, a flow-through water management systems, which allows for simultaneous entry as well as drainage of water into as well as out of the pond are imperative. The practice may be affected by the provisions of different inlets for all the ponds whereby each inlet regulates the flow of water from the supply canals to the pond or cage while each outlet controls the discharge of water outside the pond into the canal drainage. The supply and draining gates should be designed to bring water into the as well as drain water respectively out of lower levels of the ponds. This is because the water quality at the bottom of the pond tends to be contaminated faster. This is attributed to the higher levels of waste accumulation of wastes as well as subsequent decomposition at the bottom. This is ensured by the use of pumps that draw water from the various sources even at low tide so as to replenish the water regularly in the ponds so that the water supply and drainage are independent of the natural tidal fluctuations (Stenton-Dozey et al., 2021).

However, there are no hard rules on the water regulations rates for medium to high-density aquaculture. In the semi, intensive fish rearing systems it is necessary that the rate of change be at the rate of 10 % daily with an equivalent total replacement of water for at least every ten days or at least three times a month. The fish culturing systems that practice intensive management in ponds have greater organic loads at the bottom of the pond thus requires greater water exchange rates. This is especially towards the latter parts of the culture cycle whereby the animals excrete wastes that are more organic.

It is required that the ponds have aeration facilities to prevent the occurrence of anoxia, which result in fish loss of lives. This is attributed to the lowering of oxygen levels in the high-density ponds, which may result in faster rates of utilization of dissolved oxygen for respiratory activities. The situation is also caused by high rates of decomposition from the oxygen-consuming microorganisms located at the bottom. Paddle wheels or other types of aerators are thus availed in the ponds to affect the introduction of better quantities of oxygen into the water and prevent fish mortalities to a large extent. Aerators are regularly operated at periodic intervals fixed time periods during the day; however, it is mainly essential during the morning hours because of the oxygen concentration levels that are believed to be low at these moments. This results from an absence of the absence of photosynthetic, oxygen-producing activity in the pond or cage irrigation systems. Oxygen demand levels are highest towards the finish or end of the culture periods. At this time aeration, levels in the ponds or cages may be provided regularly more as compared to at the initial stages when the production was starting.

Figure 4.4: Pond preparations for freshwater farming.

Source: Krishi Jagran

The water pumps should consequently be run at faster rates for long periods of time so that the effect of water exchange is evident and useful in maintaining the fish optimum production. For efficiency purposes, pond or cage water used for fish production may be regularly sampled so as to attain measurements in the levels of dissolved oxygen, pH and salinity levels in order to determine the collective actions to be taken in ensuring that the water quality levels are to the optimum conditions for better yields in the production processes. Dissolved oxygen levels should be about above 5ppm through the pumping as well as aeration. The acidity increasing levels in the ponds and cages may be mitigated by liming of the ponds and cages. So as to ensure salinity conditions that are necessary for fish development, the ranges be maintained are 15 to 25 ppm. In the event of hot seasons that are mostly attributed to the summer seasons, there are high salinity water levels in the ponds and cages that may be diluted with freshwater from springs or wells so as to maintain the salinity levels imperative for fish growth (Sorgeloos, 2013).

In pond maintenance, regular applications of fertilizers, lime, and pesticides, as well as prevention of entry of predators, monitoring of the stock for growth rate determination as a basis of feeds and water management and regular pond maintenance, are required. Most extensive ponds including cages are fertilized regularly using organic fertilizers from cows, chicken or pig manure. However, inorganic manure may be used such as urea, ammonium phosphate or an integration of both organic and inorganic. This is so as to ensure maintenance of the plankton population in the ponds for optimum production. The application of fertilizer may be through broadcasting over the pond top surfaces of may be kept in perforated sucks and suspended from poles that are stacked at different positions along the pond periphery. Alternatively, unlike the extensive systems the intensive culture systems do not require fertilizer for they may not be naturally based with an exemption of those that grow in the plankton's feeds like the milkfish whose main diet is dependents to a large extent on the algae. The application of lime in ponds helps in maintaining the pH of water levels at alkaline or near alkaline levels. The pH levels generally should not be lower than six should. Lime may be applied by the used of broadcasting method over the pond or may be applied on the sides of the cage or pond to solve for water and soil acidity. Further, the prevention of entry of predators in the ponds and cages is also imperative. Some unwanted, as well as unwanted predatory species, survive the application of pesticides during the early stages of pond preparations while some enter the ponds via the gates screens or through cracks in the ponds. This pest may be eliminated by the application of pesticides, preferably organic.

Crabs are known predators of the shrimps, as they are carnivorous and cause damage to the pond's dike. They are best removed by the use of crabs traps situated in the ponds, as they are not usually affected by known pesticides. The gates of ponds, as well as cages, are required to be properly screened as a whole in a bid to prevent the entry of small-unwanted fish in the pond. The use of double screens at the main intake ensures that pests, as well as predators, are prevented from entering the fish farming system. Farmers are expected to monitor closely the growth and development of their fish. This is by regularly sampling the length as well as weight measurements of the fish under cultivation as a basis for determining or restating their biomass and daily feed rations. The practices are essential as it aids in making projections on the harvesting schedules and procurement of pond inputs (Skerritt et al., 2020). Basically the use of a feeding tray as a tool for monitoring stocks in the first few months as the living organisms grow in size and weight. For sampling analysis, a casting net may be used so as to acquire the sample that may be required to provide indications as to the average sizes as well as weights of the fish in the stock. Thereafter in the ponds system of farming, one can easily be able to compute as well as compute the available biomass. This indication helps in predicting the stock of fish that is surviving after initial mortalities as well as in making projections on the revenues that would be generated from the volume of harvest. The analysis may also include the estimation of daily feed that is consumed by the fish so as to come up with the profits after subtracting the production costs. Thus, in ensuring efficiency it is imperative that the costs of productions involved at each stage is recorded accurately for analysis purposes later. Additional records that might be imperative storing include data on the initial size as well as the weight of the stock of fish, the number of fry and post larvae stocked, the average body weight at each sampling group along with the feed consumption on a daily basis of the stock.

Regular upkeep of the ponds and cage gates are needed so as to identify the cracks that would if not mended lead to seepages as well as losses of stocks. Dikes in the pond are planted with grass and or vegetation covers so as to prevent soil erosion. During the harvesting of fish in a pond or cage system, the processes are done by draining the ponds or cages. Alternatively, the use of harvesting nets to catch the fish or shrimps may also be done. The latter is harvested with a bag net which is normally attached to a sluice gate as water that was used during culturing is drained out of the pond or cage. This is however done at low tide to avoid bruising of the fish. The tilapia fish is majorly harvested using seine nets, which happen mainly after the pond water is drained to half levels. This is generally recommended that it should be done the previous day before actual harvesting. In harvesting the milkfish, careful considerations are made given that the fish swims against the current. The method used in catching this type of fish is known as the "pasulang" from the Philippines. The method involves draining about 85 to 90 % of the pond water during low tide while at the same time allowing in the water at the incoming high tide. This is so as the fish could swim against the current via the tertiary gate into the catching pond or cage. It follows that the gates of the pond or cages are closed as soon as a big quantity of fish stock is impounded. Lastly, the fish in the catching pond are harvested by seining and handpicking.

Farmers have come up with ways of minimizing their costs while engaging with pond or cage fish production systems. Some of the methods used in minimizing costs at the farm level involve the application of integrated farming where farmers practice fish culture in collaboration with other farming activities such as the farming of crops. These crops include rice and vegetables. Alternatively, farmers keep animals such as pigs, chickens, ducks and cows. Inefficiency in the production of fish is achieved from the overall production efficiency levels at the farm as a whole. This occurs when the wastes or byproducts of one farming component is used as an input in another farming component. For instance in the fish pond, poultry manure or other manures generated from pig, cow, ducks, etc. as well as the vegetable wastes are used as fertilizer for the fish in the pond. The vegetables can also be grown so as towered to the fish, which is more cost-efficient. In Africa, integrated farming in fish is mostly seen in the rice fields where fish are reared along with rice in the same field. The combination of different activities with the rearing of fish has been reported to be economical as well as varying with the different combinations of alternatives.

Figure 4.5: Smart Farm -Caged Fish farming system.

For example, the fish yields generated from a fish enterprise that is mixed with ducks are around 2000 to 4000 kg/ha/y while that which is mixed with pigs generates a fish yield of about 8500 to 8900 kg/ha/y and an enterprise that is combined with poultry is generated about 3600 to 4900kg/ha/y. Past studies have reported that the combinations are of economic benefits in that they involve minimal investments in the fish industry. The use of integrated fish farming in the recent times has however been constrained by the widespread use of pesticides in most countries which are not good for the growth and the development of the fish. The cage culture entails fish rearing within fixed or floating nets supported by structures or systems made of bamboo wood or metal that are mostly set in sheltered, shallow portions of lakes, Bays Rivers and estuaries. The system can be assessed independently from two countries, one is in Kampuchesa where the fishermen in the great lake regions used to keep commercial fishes in bamboo cages and baskets, especially fish of the Clarias species. The other country is Indonesia where bamboo cages have been used in culturing *Leptobarbus hoeveni* fry since 1922. Today the use of cage fishing system has spread to over 35 countries across Europe, Asia, Africa as well as the Americas (Shaalan et al., 2018).

4.4 INTEGRATED RECYCLING SYSTEMS

This is seen as the largest scale method for the raising of pure fish on a farm. This method uses large plastic tanks that are placed inside a greenhouse. In this method, there are hydroponic pipes that are placed close to the plastic tanks. The water that has been used up in the water tanks is then circulated to the hydroponic beds, this is where the fish feed, and fish waste go to provide nutrients to the plant crops that have been grown in the hydroponic beds. The most common plants that are usually planted in the hydroponic beds are small plants and herbs such as basil and parsley.

In most instances, there is a shortage of water especially freshwater that is required in the farm. The rearing of fish which is mostly referred to as aquaculture or pisciculture, a huge amount of water is required and it can use up to a million gallons of water per acre. This means that water purification systems will allow the reuse and/or recycling of water that is already available. In the largest scale fish farms, they tend to see systems that were derived from the New Alchemy Institute in the 1970s. In this process a large fish tank majorly it is made of plastic is placed inside a greenhouse. The hydroponic bed can then be placed above, near or between the plastic water tanks.

Figure 4.6: Integrated recycling system. The fish are kept in tanks and pipes are placed above or on the side of the tank with hydroponic medium and plants can be grown.

Source: Oxfarm

When fish such as tilapia or catfish are raised in these tanks, they are known to feed on the algae which will naturally grow inside and on top of the water when they are properly fertilized. The tank water that the fish is using is then circulated to the hydroponic beds, where the nutrients that are in the water are used to nitrify the plants that can be sold commercially. The nutrients that will be taken up by the plants will be derived from the waste from the fish or the fish excreta, it also comes from the commercial feeds that are fed to the fish but in most instances, the fish will not consume all the feed that has been provided for them in the tank (Shah et al.,2018).

As the water, passes through the hydroponic beds there are microorganisms that have been carefully cultured so that they can be used to convert ammonia to form nitrates. Ammonia is poisonous and harmful to plants hence they cannot be readily taken up by plants but when converted to nitrates they can then be readily taken up by plants through the roots and be utilized by the plants. The plants can then be fertilized by the nitrates and phosphates that are now readily available. The other wastes that are useless to the plans are strained out by the hydroponic media, which also acts as a pebble bed filter and is aerated.

In his type of system if it is properly tuned it is able to produce more edible protein per unit area than any other. There are a wide variety of plants that do well on the hydroponics bed. Most of the individuals that use this system grow herbs such as basil and parsley, which can fetch very good prices in the market even in small quantities throughout the year. Most users that consume these herbs are restaurants and wholesalers.

Figure 4.7: In this type of system the tanks are inside a greenhouse, which means they can be used in both tropical and temperate climates.

Source: MSD AQUACULTURE

As seen, earlier this system is fully placed inside an enclosed system i.e. a greenhouse and this means that it can adapt to all temperate climates and this means it can adapt to tropical climates. In this instance, the biggest environmental impact is how to discharge water that is salted in order to maintain the fishes' electrolyte balance. This is the case because, at most times, water that has been recycled and taken back to the tank generally does not contain salt as most of it has been taken up by the plants. This then means that most of the current fish growers and individuals that use integrated recycling systems are using proprietary tricks, which is enabling them to keep the fish healthy and to also reduce the cost that may incur from buying the salt that is needed for the fish and the expense of getting a permit to discharge the water. Authorities involved in veterinary services are speculating that some of the farmers and companies involved are using ultraviolet ozone disinfectant systems, which in most cases are used in the rearing of ornamental fish. These systems as they speculate are used in keeping the tilapia healthy with the water that has been recirculated.

Most of the companies that are regarded as large and well capitalized in this area have mostly failed if not all. The main problem to them is managing the biology and markets which at most times to navigate these two at the same time is very complicated. One of the developments that is seen to be a thing for the future is integrating the recycling systems with urban farming and this is already being tried in Sweden by the green fish initiative (Wurts, 2000).

In this system, there is a balanced ecosystem where there is no waste. In this instance, it can be noticed that the waste that has been produced by the fish is utilized by the plants. The plants then produce oxygen that can be used by the fish as they breathe in. This brings a balance, as there will be no accumulation of waste in the pond or in water.

There is efficient waste utilization for different cultural practices for the production of fish. The waste that is produced from fish production is much and can be expensive even to purify. The waste is utilized naturally by other living organisms i.e. plants.

This system helps in the reduction of additional costs such as supplementary feeding and fertilization. The plants that are grown using hydroponics do not require nutrients as they readily get nutrients for the microorganisms that convert ammonia to nitrogen.

There are more employment opportunities. In this instance, there are individuals that will be employed to deal with the fish production while other individuals will be required to take care of the plant's sections. This increases the job opportunities available and in turn, people living around can get employment to sustain themselves.

This system is seen to reduce the input but is seen to increase the output. The farmer or company is able to harvest on two ventures i.e. the fish and the plants. This means they are to harvest all year round because they harvest fish and herbs almost at different times which means their general income in the farm and the usage is low, as the input required has been reduced.

A farm such as this has the potential to increase production and the socio-economic status of the surrounding area. In most instances, these farms due to their largeness will employ a large number of people in the locality of the company and they will also improve the amenities around them so as to enable their employees to live comfortably so as to increase their production and to reduce losses such as time.

In this type of system where the water is, being recycled there will be reduced amounts of dissolved oxygen. With fish being kept in high density in the tanks and with their feeds containing large amounts of protein will reduce the amounts of oxygen that will be available for the fish to breathe. In the water, oxygen is consumed by fish as well as during the breakdown of waste and to an extent the water is less aerated as it is pumped back to the tank from the aquaculture, which means they have not been exposed to air. At this point, the farmer will have to add oxygen pumps, which can then pump dissolved oxygen into the water, as it is most important and if neglected it can lead to the death of the fish.

Figure 4.8: Oxygen pumps are used to supply oxygen into the plastic fish tanks and add to the dissolved oxygen in the water.

Also, the water has to be filtered and it is done both mechanically and biologically. During mechanical filtration, it is used in order to remove the solids from the system. These types of filters will mostly include separators, screens or granular filtration. On the other hand, biological filtration will include the necessary bacteria that will help in the removal of unwanted nutrients and conversion of some if possible. The filtration will depend on the availability of bacteria as well as the number of bacteria that are available. Bacteria are microscopic and hence can disappear and reappear. This will affect the quality of water moving back into the tank.

This venture requires a lot of capital in order to start, as one needs to get permits depending on the country, the construction of the greenhouse, the purchase of the plastic tanks, and purchase of pipes for the hydroponics and the hydroponics medium. The owner will also have to employ skilled and unskilled labor, which will increase the wage bill, especially for starters.

Integrated recycling systems are especially good for the conservation of the environment and it also promotes food security that are some of the millennium development goals. This venture needs to be adopted so as to improve the environment and also provide food and employment to many (Yang et al., 2020).

4.5 CLASSIC FRY FARMING

It is also referred to as the flow-through system. This is the process whereby fish for example trout and majorly other sport fish are raised from eggs to fry or fingerlings, they are then tracked to streams and released. In most instances, the fry is often raised in long, shallow, concrete tanks and fed with water that is fresh from the stream. The fry or fingerlings are usually fed commercial foods i.e. pellets. This system has been used for several years although not as efficient as the integrated recycling system but it is a lot simpler and does not require a lot of skills to establish and run.

In this type of system the water usually flows through the system that allows for waste treatment where the water leaves and is not recirculated into the system. This means the water does not require any treatment. The water that flows into this system needs to be of the required temperature so as not to affect the fish and should match the optimal temperature of the fish so as to ensure there is success for the fish and to avoid unnecessary stress to the fish. Oxygen is brought in by the fresh water and hence is readily available to the fry. The water containing carbon IV oxide flows downstream and does not come in contact with the fish again.

This system requires that there is a constant flow of water throughout the year. This means the stream where the system is to be established the stream has to be a continuously flowing river the whole year round and is not affected by dry seasons.

Figure 4.9: The European eel, it is deemed to be becoming extinct but is now being raised by agriculturalists using the classic fry system.

Source: Non-indigenous aquatic species USG

The European eel i.e. the *Anguilla anguilla*is being threatened by extinction due to the fact that the glass eels are being caught in large numbers by the Spanish fishermen and there is also overfishing of the adult eel by Dutch fishermen in Ijsselmeer in Holland. This has aqua culturalists to procure limited supplies of glass eels, which are the juvenile stages of the European eel, which are known to swim north from the Sargasso Sea, which is known to be their breeding grounds to their farms where they can be protected from the fishermen and this can help in avoiding their extinction. The European eel larvae are known to survive for several weeks but it has yet to be proved that they can stay their full life cycle while in captivity.

This system is mostly used for the culture of rainbow trout *Oncorhynchus mykiss* and other examples of salmonids in the United States. The other examples of cold-water fish that use this system include the brook trout *Salvelinus fontinalis* and the brown trout *Salmo truta*. This system is also used for limited production of some warm water fish, which include the catfish *Ictalurus* spp. and tilapia *Orechromis* spp. New species that have been considered and being tried recently include the yellow perch *Perca flavescens*, the hybrid strip bass *Morena* spp. and some species of the sturgeon *Acipenser* spp.

There are several constraints and risks that are involved in the using of the flow-through system. Large amounts of water are required. Due to the fact that water is needed in large quantities because the water flows freely through the system, it means the water source and supply needs to have consistent amounts of water and in case of a little amount of water it may lead to failure of the system and the fries will eventually die (Yang et al., 2020).

The fries require expensive feeds. The fries are delicate and they need expensive commercial feeds that are rich in proteins so as to ensure they grow healthy and strong so as to ensure that they can have energy and stamina to depend on themselves when they will be eventually released into the free water.

These fish are vulnerable to catastrophes. These fish are raised in the open and they are vulnerable to catastrophes such as flooding and drought. In case of too much rain, the level of water will rise and the fry will be carried away by the excess water. In some other instances, there can be mudslides which the river and/or stream will carry away the silt and may deposit them in the system which will affect the fish. In instances where there is a lack of rain in the region, the water levels decrease and in turn, the fry and the system do not get sufficient water.

The need for one to acquire a permit. In most countries, for one to be involved in classic fry farming permits need to be acquired because the individual will be using natural resources such as streams and springs. This means an individual has to continually comply with government set guidelines on the protection of natural resources and not anything carried out should influence the habitat or ecosystem found close to the system.

The quality of water in this type of system is very crucial. The water quality that is entering, the system needs to be pure, clean, and free from minerals. The water needs to contain dissolved oxygen, which can be readily available for the fish. The salt content needs to be at an optimal level to ensure good conditions for fish viability. It should not contain minerals or chemicals so that it may not be consumed by the fish.

The water needs to be from clean sources such as forests because the water from forests does not contain any chemicals because it is yet to interact with humans. An individual cannot use water from urban areas, which contains a lot of chemicals example from industries and from surface runoff. The water is known to be quite dirty and cannot be used for fry rearing.

In this system, individualism has to consider the topography of the place so as to ensure the free flow of water. In this system because of the fact that it is a flow through the area has to be gently sloping, which will then ensure that the water will flow without stopping. In a generally flat area, the water will hardly flow and this will lead to accumulation of waste and this will lead to toxicity of the water and lead to the death of fish and failure of the system (Zhang et al., 2015).

Although there might be setbacks and some risks involved this system is quite useful to the farmer. This system allows for closer monitoring of the growth and mortality of the fry. Due to the fact that the water flows freely, this means the fry are easily observable and, in many instances, they are less affected by predators and/or diseases.

Figure 4.10: Feeding in a flow-through system. It is easy and efficient hence does not require a lot of skills.

Source: farms.com

This system is easy to maintain. There is little or no skill required to run this type of system, which means it is cheap and easy to maintain. The flow-through system means the individuals running the system do not need to have high-level knowledge, as what is required in this system is mostly feeding and taking inventory to ensure the fish are living in good conditions.

Feeding of the fry is quite easy and efficient the simplicity is the feed is just placed on the water surface and the fry can easily reach the feed. The feed is also only given to the fry at specific times during the day. This system can be used to raise a large number of fish as compared to other systems. This is because this system allows for the raising of fingerlings up to a certain age then released which means within a short period of time another batch of fingerlings can then be raised.

It is useful for the restocking of fish in the natural environment. Due to the fact that the fish raised in this system are raised and released to the natural habitat, it means there are less chances of extinction of some fish in the ecosystem. This also ensures there is fish in lakes and seas as in such instances there are fishermen that are overfishing and reducing fish numbers that can lead to an imbalance in the ecosystem to affect the habitat for quite some time.

Chapter 5

Defining Urban Aquaculture

CONTENTS

5.1 INTRODUCTION

It would be helpful to explore a basic description of this industry or more precisely a group of operations, before examining the current situation, prospects and restrictions for urban aquaculture, as is provided below. First of all, it is value taking into account the urban landscape in which aquaculture is believed to have three main criteria in that review, as stated by some recently developed urban specialists or more adequately communities. First, the urbanization of a particular place is related to changing demographics, which is an increase in population size and density.

Second, the economy pushes toward a labor force fundamentally occupied with nonagricultural exercises. Thirdly, those living in urban regions have a social-mental awareness of what this implies. Notwithstanding, segment and monetary cycles bringing about urbanization don't happen equitably around urban regions, and numerous components impact the rate and degree of urbanization. Besides, urbanization isn't in every case straightforwardly connected with improvement around prior urban places. The social-mental component of urbanization implies that through cycles like the exchange of settlements and non-income assets from provincial urban transients; dispersion of consumerist and urban thoughts and modes of conduct; support of returning travelers in local area dynamic, even rustic networks may go through urbanization.

Figure 5.1: Aquaculture.

Source: https://en.wikipedia.org/wiki/Urban_aquaculture#/media/File:Tiang_Sabah_Fish-aquaculture-01.jpg

As per recent studies, peri-urban conditions and networks share numerous aspects with those which are viewed as urban; typically the progress from urban to periurban to country networks is viewed as a continuum. In this survey we consider that aquaculture exercises embraced in both urban and peri-urban settings share numerous attributes, in any case, we additionally desire to exhibit that as networks or conditions become more urban in nature, then, at that point the administration of aquaculture should turn out to be more concentrated, in any case, exemptions and impediments exist. Now it is helpful to think about the thing is implied while thinking about aquaculture.

Aquaculture might be characterized as "the cultivating of aquatic natural entities, including fish, mollusks, cephalopods and marine plants. Cultivating infers some type of mediation in the raising cycle to upgrade production, like the standard stocking, taking care of, security from predators, and so forth cultivating likewise infers individual or corporate responsibility for stock being developed. For factual purposes, oceanic life forms that are gathered by an individual or corporate body that has claimed them all through their raising period add to aquaculture, while fish natural entities that are exploitable by general society as normal property assets, with or without suitable licenses, are the reap of fisheries". Another set of experts though, conversely recommend that in case there is mediation to build yields or potentially responsible for or controls on admittance to and benefits accumulating from intercessions that this ought to be classed as a type of culture (Sandersen & Kvalvik, 2015). From the FAO definition, it ought to be noticed that sea-going macrophytes, developed in numerous Asian districts comprise a significant aquaculture action, notwithstanding, this production isn't recorded in the FAO measurements.

Figure 5.2: Fish farming in Hanoi (photo by Huub Ruijgrok, taken from video Urban Agriculture).

Source: https://www.researchgate.net/publication/237452033_Urban_Aquaculture

Aquaculture farming, characterized as "a framework wherein adolescent fish are delivered to develop, unprotected, on normal food varieties in marine waters from which they are gathered at attractive size" may likewise be considered inside the FAO definition where possession can be held or moved. The FAO definition seems to prohibit groups other than people and corporate bodies from participating in aquaculture, notwithstanding, families, families, networks, co-agents and governments all participate in aquaculture exercises.

This is of specific significance with regards to urban aquaculture where the fair administration of scant natural resourcesis probably going to rely on local area-based associations holding possession, or if nothing else the option to misuse oceanic plants and animals refined utilizing normal property assets.

Thusly, with the end goal of this review, urban aquaculture might be characterized as the act of aquaculture happening in urban settings, or regions subject to urbanization, joining by definition, peri-urban regions. In the accompanying area, various ways to deal with urban aquaculture are inspected as for the species refined and power of development.

5.2 ATTRIBUTES OF AQUACULTURE PROJECTS

Aquaculture involves controlled cultivating or development of life forms in salt, bitter or new water with the end goal of food production. Aquaculture produce is additionally effectively utilized in different associations, for example, drug or restorative ventures.

This booklet for starting ecological appraisal covers aquaculture projects comprising of cultivating/development and gathering of fish, shellfish, mollusks (shells, snails) or ocean growth.

Cultivating crocodiles and turtles are additional types of aquaculture, yet won't be managed in this booklet. Notwithstanding the aquaculture farm accordingly, an undertaking may incorporate offices for the production of fry and feed. Some aquaculture activities may likewise require offices for the preparation and circulation of the aquaculture items. These can be isolated undertakings or sub-ventures of significant aquacultural speculation. An aquacultural venture may likewise involve measures as far as preparing, research, enactment, counseling, advertising, organization and foundation of establishments.

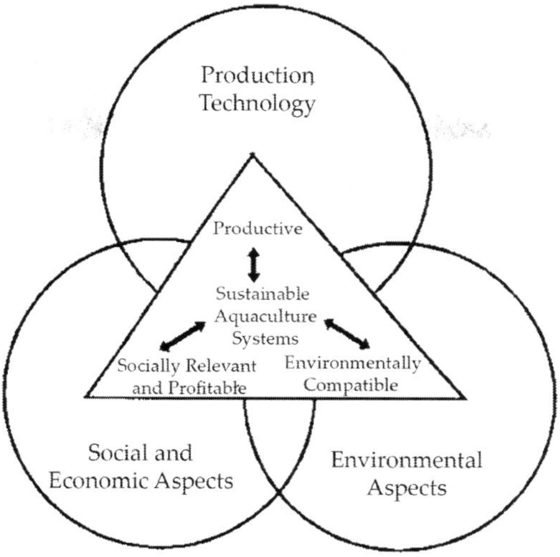

Figure 5.3: The development of sustainable rural aquaculture systems involves consideration of production technology, social and economic aspects, and environmental aspects.

Source: http://www.fao.org/3/x6941e/x6941e07.htm

The decision of innovation, scale and power of development will change an incredible arrangement and will be conclusive for the ecological and socio-social effects of aquaculture. Aquaculture undertakings may change as to loading of living beings, arrangement of the production regions and the level of taking care of, preparation and drug. The farm living beings can be restocked at various phases of advancement. For instance, they might be wild fry that has been gathered for additional development (Rosas et al., 2019). In different cases, the fry are farm-developed. Import of fry from different locale or nations is another other option. Setting up the production regions may include making fenced-in areas in straights, tidal ponds, lakes, lakes or man-made lakes/repositories, the structure of artificial banks and bowls, and the utilization of enclosures. A pen is an open, free gliding net appended to a structure for use in both freshwater and saltwater conditions.

It is entirely expected to recognize extensive, semi-intensive and intensive systems, contingent upon the level of taking care of and treatment in the homesteads. Extensive cultivating includes no taking care of and no, or negligible, preparation, semi-intensive some valuable taking care of and treatment, and serious consistent inventory of feed and composts. The more intensive the production, the higher is the density of farm animals in the cultivating medium. In concentrated cultivating systems, focused on a high production, the density of homestead life forms is regularly exceptionally high. Additionally, extensive systems depend on the development of only one animal group (monoculture). The climate consequently turns out to be altogether different from a characteristic one, and the homestead animals might be presented to pressure and a more noteworthy defenselessness to disease. Appropriately, the requirement for medicine will increment concerning the level of rise. In extensive and semi-intensive cultivating there will seldom be a requirement for a prescription, though medications like anti-microbials, fungicides and parasiticides are important in intense systems.

In non-industrial nations, there are intensive, semi-intensive as well as extensive cultivating systems. Extensive and semi-intensive systems are right now the most well-known. In a few spots, notwithstanding, there is developing interest in the advancement of intense systems. As a rule, extensive and semi-intense aquaculture can be supposed to be all the more effectively versatile to a specific climate, and the potential natural issues will be less and less genuine than those brought about by intensive aquaculture.

Extensive and semi-intensive cultivating systems are regularly in fact basic, however can simultaneously be naturally confounded. A few animal varieties are regularly developed all the while (polyculture), misusing distinctive biological specialties. In the wide-open semi-intense systems frequently join horticulture and aquaculture. In this manner, rural decline, excrement from homegrown animals and so forth are used in fish-lake production. Fish cultivating can likewise be founded on different supplies of supplements like sewage and septic waste. Extensive and semi-escalated systems are by and large work-saving and less capital-intensive, many have been worked for a few ages and are balanced to nearby conditions. The improvement of extensive and semi-escalated systems is regularly designed towards production for nearby utilization. By and large, accordingly, no significant improvement of handling and conveyance systems is required.

The development of mollusks and kelp happens extensively. Mollusks, for instance, will exploit the nearby production offices by separating particles from the waters. Such a farm may in this manner, somewhat, fill in as a purifier plant, as the shells channel little natural particles. The life forms to be developed are restocked or planted, for instance on instruments where the living beings can attach themselves.

Figure 5.4: Hydrotech belt filter used for dewatering the sludge.

Source: https://eurofishmagazine.com/sections/aquaculture/item/254-guide-to-recirculation-aquaculture-chapter-6

Verifiably intense cultivating systems are somewhat nascent. They have been presently advancing since the turn of the century, however, have gotten an amazing force somewhat recently. It is predatory (carnivore) types of fish and shellfish that are being developed intensely.

The prerequisites as to control and the board of the taking care of, prescription, water quality and volume are higher than those applying to different systems. Further, the clean necessities are high, since the danger of contamination from medications, supplements and natural matter is high. Normal support and cleaning of the homesteads are vital. Intensive cultivating frequently requires an extensive capital venture and the accessibility of a prepared and forever utilized labor force. In any case, the types of cultivating are not particularly work- intense. They frequently include an extraordinary expansion underway and will, by and large, require the foundation of offices for gathering, preparing and dissemination of the homestead items.

The advancement of gathering, handling and dispersion mechanical assembly for aquaculture production requires land and a steady inventory of energy (oil and power) and excellent water. This particularly applies to present-day techniques for protection like icing and freezing, however conventional strategies for treatment, for example, smoking, additionally rely upon a guaranteed supply of energy. Less energy-consuming techniques, for example, drying and salting, might be significant other options. Expanded interest in aquaculture, also, requires an adequate stock of building materials, for example, wood for the structure of pens or pens (Naylor et al., 2021).

The treatment of a lot of farm items in a confined region can cause issues with the treatment of waste. Waste from for example the dealing with and preparing of fish comprises of effectively decomposable natural matter which represents somewhere in the range of 5 and 30 percent, and in exceptional cases up to 70 percent, of the crude material, contingent upon the level of handling. This waste is an asset that may yield a progression of important items, for example, feed, fish supper, fish oil, enormous gas, manures and so forth, or structure a reason for the production of more particular natural items.

Figure 5.5: Combined intensive-extensive fish farming systems in Hungary. The number of opportunities seems unlimited.

Source: https://eurofishmagazine.com/sections/aquaculture/item/254-guide-to-recirculation-aquaculture-chapter-6

As of now referenced, a significant advancement of aquaculture is probably going to prompt the improvement of a few related capacities. In such cases, one can likewise counsel the agendas or potentially booklets for starting ecological appraisal of industry, transport, improvement of thickly populated or urban regions and so on.

If the aquaculture project contains preparation as well as the advancement of organization and foundations, natural viewpoints and the topic of ecological capability ought to be given cautious thought.

5.3 APPROACHES TO MODERN AQUACULTURE

Considering the scope of species delivered in aquaculture systems, this survey will cover the most critical and inescapable gatherings created in urban settings, including fish plants. With numerous urban places situated in marine regions, note that urban aquaculture, albeit likely overwhelmed by production in freshwater, may likewise include aquaculture in brackish water and marine conditions. Considering momentarily the circulation of aquaculture internationally, the FAO's information base expresses that in the 90s, Asia represented 89% of aquaculture production, Europe represented 6.3%, North and South America, 2.1 and 1.9 percent, separately, and Africa and Oceania under 1% each.

The circulation of contextual analyses introduced in this audit generally reflects the present circumstance, with most records of urban aquaculture coming from Asia, and a generously lesser number from different areas. Further to portraying urban aquaculture as dependent on the spot, species and climate concerned, the audit introduced here will endeavor to exhibit that the power at which urban aquaculture is overseen additionally differs in light of outside pressing factors and motivators for producers. Conjuring aquaculture systems typology that has been developed we can all the more likely depict the idea of aquaculture happening in urban regions. Extensive aquaculture is portrayed by the reliance of stock on regular food; semi-intensive production includes compost applications to improve normal food production and additionally the arrangement of beneficial feed, which is typically low in protein; culture in intensive systems depends only on an outside supply of high-protein (>20%) feed.

Essentially, nonetheless, these qualifications can become obscured. Numerous peri-urban culture systems depend on drainage or the directing of prolific run-off or 'dark water' into water bodies. These information sources might be pretty much unregulated albeit the gather of items as fish or plants might be profoundly overseen. In numerous rustic regions where requests and markets for oceanic items are restricted, it isn't unexpected for producers to embrace extensive aquaculture practices or semi-intensive methodologies, however with just choice or limited mediations. Frequently aquaculture in such settings is essential to guarantee family food security, however of restricted potential for creating generous off-farm pay. In specific circumstances, it might likewise be feasible for networks to depend on wild catch fisheries, or if nothing else upgraded fisheries, to meet both their requirements and neighborhood interest. At times aquaculture can turn into a system to diminish or alleviate chances from dubious regular fisheries; loading fish turns out to be more well known in pieces of Cambodia when the downpours, and fish yields from the ricefields, are less plentiful. Nonetheless, as marine assets in even the remotest settings become privatized and access becomes confined, numerous country networks are compelled to do without conventional fish food varieties or to put resources into aquaculture, but of an extensive or restricted semi-intensive level.

In peri-urban regions, admittance to bigger business sectors and more predictable and dependable interest, average producers are bound to put resources into a more extensive scope of semi-intensive administration procedures. With the more noteworthy accessibility and centralization of homegrown waste, specifical wastewater from urban seepage systems, and results from food preparing and promoting, average producers are additionally ready to adventure such assets, decreasing their consumption on composts and feeds. Obvious endowments to peri-urban farmers as waste assets seem to offer them a huge benefit over producers with restricted admittance to such production upgrading inputs. In chose models, admittance to such assets has to lead to open development in peri-urban aquaculture and far and wide advantages for producers. A relevant illustration of this is the tilapia seed producers using sewage close to Ho Chi Minh City that enjoy an important cutthroat upper hand over different producers on the Mekong Delta, even after transportation costs.

Figure 5.6: Urban aquaculture and agriculture.

Source: https://earth.org/urban-agriculture/

Be that as it may, despite the advantages of being found near urban business sectors and having the option to get to waste assets, there are potential imperatives related to undertaking aquaculture in peri-urban regions. While rehearsing extensive aquaculture, producers are regularly unfit to apply command over the overall hydrology, while farmers overseeing huge water areas for semi-intensive aquaculture may think that it's hard to control all inflows and releases. In such conditions, the receptiveness of the way of life framework may permit toxins, predators and illnesses to enter and supplements, food assets and stock to runoff. The actual transparency of numerous extensive and semi-intensive aquaculture systems likewise implies there are chances from airborne contamination, especially agrochemical splash debris, predators like piscivorous birds, and burglary by poachers. This last issue is regularly exacerbated by the settlement examples of destitute individuals in periurban regions who frequently site their homes on the outskirts of water bodies (Milstein, 2005). Taking into account that numerous requirements to production in extensive and semi-intensive aquaculture systems are outside the ability to control the farmer, one potential administration procedure to restrict the dangers presented by such risks is to limit the receptiveness of the way of life framework. Nonetheless, as both extensive and semi-intensive production exercises rely upon ecological labor and products to enhance contributions from the farmer, confining the transparency of culture systems requires the farmer to either lessen production or to make up for the misfortune by expanding inputs. Outer constraints that lead farmers to receive more intensive production procedures might be considered as 'driving instruments'; moving to more intensive production regularly requests monetary speculation, obtaining more responsive and gifted administration strategies, and more prominent exertion committed to observing system soundness, systems working and likely outside risks.

Subsequently, albeit to a great extent used to sort aquaculture systems dependent on creative power and the board requests the grouping is useful in demonstrating the level of control and reconnaissance administrators can work out, both basically and as far as the resources whereupon they need to draw. The change from extensive to semi-intensive might be credited to different components, notwithstanding, more prominent interest from business sectors, joined with further developed showcasing channels, frequently establishes a significant driver for heightening. Control of assets, more admittance to production improving contributions, for instance, waste assets, food handling side-effects and credit to buy extra seed, feed and work can likewise invigorate increase. In urban settings the progress from semi-intensive to intensive production seems, by all accounts, to be driven by generally monetary contemplations and expanded rivalry for assets, specifically land, yet additionally strong natural and wastewater assets, work, credit and markets. Strengthening likewise seems to offer producers more noteworthy control, empowering them to more readily protect and upgrade the nature of items, tending to concerns communicated by consumers in regards to conceivable security risks. Notwithstanding the upper hand related to strengthening, a few hindrances to such progress can be recognized; exchange expenses might be high, while restricted admittance to information, preparing, credit, markets and institutional help limit the choices and openings accessible to producers.

Winning qualities of urban aquaculture surveys ongoing records of urban aquaculture, evaluate the scale, dispersion and relative significance of urban aquaculture and depicts the predominant administration attributes of extensive, semi-intensive and intensive production systems.

5.4 INTENSIVE URBAN AQUACULTURE

While ostensibly past the focal point of this appraisal, intensely controlled aquaculture adventures in urban communities are being made by trailblazers in various countries, regardless, such firms are fundamentally bound to North America and Europe. However, undeniably less land might be required per unit of the item for escalated instead of semi-intensive production units speculation costs related to building intensive systems are somewhat costly. Intense systems profit with expanded control by labourers over the framework's working taking care of better factors like nature of water, feed supply and stock administration.

More intensive, less open systems likewise offer the maker more noteworthy authority over open, animal and environmental security risks. Nonetheless, because of the high capital and working expenses of escalated systems, by and large, it is simply plausible to create high-value items, which are frequently bound for expert business sectors. In North America and Europe, escalated urban aquaculture systems have been utilized to create high-value fish like tilapia, ocean bass and eels.

Figure 5.7: Overview of urban aquaculture.

Source: https://edis.ifas.ufl.edu/publication/FA217

5.5 EXTENSIVE AQUACULTURE

Extensive aquaculture incorporates systems of culture and raising in which human intercession is focused on the propagation of the stock, notwithstanding. When contrasted with fishing, these systems make it conceivable to specifically build the production of the species generally valuable for man's food (salmon, clams, and so on), industry (algaculture), or delight (sport fishing, pearl culture, and so forth) They utilize the planktonic or benthic production of the water masses, in this way streamlining on the treatment of the reproducing climate (semi-intensive systems), or the swelling of the stock (intensive systems).

Extensive practices are not new. In Europe, the stuffing of mussels on stakes traces back to the fourteenth century. The benefit of catching altercation as a solution for the overexploitation of normal shellfish beds was acknowledged as right on time as the second 50% of the nineteenth century. The benefit of relocating juvenile fish from their nurseries in the south of the North Sea to less densely populated tidal ponds or ocean buffering zones, toward the start of this century. The authority of artificial multiplication invigorated excitement for ocean farming: the principal arrivals of juvenile salmon from incubation facilities date from the middle of the last century (Europe, Pacific shorelines of America and Japan), the primary cod incubator was inherent Norway in 1884, and the main lobster incubation facility in 1891 on the Atlantic bank of Canada (Mazur & Curtis, 2008).

Information about the real production of a few extensive systems is as yet restricted. While the production of stationary species (shellfish and green growth), whose stocks are geologically particular from the normal beds and by and large possessed by an individual or a corporate body, is genuinely notable, the way of life of vagile species (predominantly fish and cephalopods) whose stocks are misused all things considered simultaneously as wild stocks and whose catches are not recognized from those of the regular population, isn't at all notable. This vulnerability has driven the FAO to reject from its aquacultural insights the production of by and large harmed stocks. This choice prompts the incorporation in the fisheries measurements of production coming about because of the arrival of juvenile fish into watercourses, lakes, seaside waters, and the sea, even though this production is exceptionally huge for a few animal types, specifically salmon.

Algaculture and shellfish culture, with a production of 7,7 and 7,3 million tons individually in 1996, put together for the most part concerning the extensive modes, address half of world aquacultural production as characterized by the FAO. European shellfish production surpasses 750,000 tons. In Japan, the production of salmon by farming has expanded ten times in the course of recent years, to surpass 200,000 tons today. In 1994, the quantity of relocating fish in the waterways of Hokkaido was multiple times the past greatest saw in the whole history of fishing.

Figure 5.8: The USDA National Aquaculture Research Center in Stuttgart, Arkansas.

Source: https://edis.ifas.ufl.edu/publication/FA217

From the biological angle, the potential for the development of extensive aquaculture seems impressive. Speculations on the proliferation methodologies of fish populations, the plentifulness of the normal vacillations of wild populations, and the productions accomplished in certain extensive cultivating systems demonstrate that sea-going environment could uphold stocks far greater than the normal biomass of the wild stocks. Besides, since the volume of the marine biological systems damaged by these strategies is still somewhat little, it is reasonable that this method of production ought to animate exploration premium. The number of marine species considered rose from 150 out of 1984 to 250 in 1994.

The outcomes have been frustrating notwithstanding. Lately, barely any R&D programs have prompted any critical production. Shellfish culture is encountering overloading issues that are bringing about the crumbling of the specialized and financial loads of the endeavors concerned. In the work to lessen these uneven characters and deal with the contentions which result from the contending employments of seafront biological systems, conventional techniques for guidelines have been demonstrated ineffectual.

The possibilities for extensive aquaculture can be assessed assessing the condition of the area. Since the 70s, the yearly rate of expansion in world halieutic production has continually been falling - in value terms significantly more quickly than in weight -, to become negative throughout this decade. The justification for this damage is the damage of wild stocks and the inadequacy of the traditional strategies for managing admittance to fisheries whose opportunities for extension are exhausted. The aquaculture improvement arrangements presented toward the finish of the 60s to take over from traditional fishing began to prove to be fruitful during the 80s. For as long as fifteen years world aquacultural production has been developing at a yearly rate of 8.5 %. It presently represents over a fourth of absolute production. Progress remains unevenly appropriated anyway as far as development systems, production locales, items and markets. Notwithstanding the stupendous accomplishment of salmon cultivating in northern Europe or prawn culture, traditional systems (small shellfish farms, freshwater fish farms, and algaculture, predominantly in Asia) stay the primary recipients of the new development.

Because of the stagnation of fishing profits, European nations overall are confronted with both a quantitative and subjective deficiency of marine items. In 1996, they imported over US$ 21 billion of marine items, and the shortage added up to US$ 6 billion. For the European Union alone, it was

US$ 8 billion - and separately US$ 16 billion and US$ 4 billion for Japan and the USA (Maulu et al., 2019). Just Norway, Iceland, Russia, Denmark, Ireland and Ukraine are, in sliding request, net exporters in value terms. As in the greater part of the rich nations, the gap was chiefly in top-notch items (fish, shellfish, fresh, filets and frozen). In these conditions, public strategies should keep on supporting the extension of aquaculture and the broadening of production systems, with the need being given to those worried about types of high business value.

The marine climate is appropriate for the examination of the ramifications and conceivable outcomes of extensive aquaculture. Given their enormous measurements, and because, even in marine regions, the oceans and seas experience normal less pressing factors than inland waters, marine biological systems offer better possibilities. The size of the maritime masses uncovers better the issues that the ease of the environment and the versatility of the stocks address for their domestication. Any development in the law of the ocean reveals better insight into the job of the organizations in the misuse of natural resources and the justification of their uses.

5.6 THE PRIMARY EXTENSIVE SYSTEMS

Specialized heightening makes it conceivable to surpass the usefulness furthest reaches of the wild stocks. The interaction comprises of expanding the specialized controls of physiological capacities like propagation, specific phases of the existence cycle (eggs and hatchlings, adolescents, grown-ups, reproducers), just as their development environments. This domestication happens through discrete leaps: among fishing and controlled development there is a scope of systems that can be arranged in four gatherings:

i. those which concern generation,

ii. those which likewise mediate in the richness of the development environment,

iii. those in which the stock is additionally taken care of, and

iv lastly, those where the wastes are additionally reused. Not the entirety of the possible systems are practical, and the most intensive are not the most productive.

While specialized intensification diminishes the job of the normal asset, it doesn't dispense with the reliance on production versus the climate. It changes just the idea of the obliging components and the limits that they force on production. In fishing, it is the tertiary production - that of the halieutic populations - which is the restricting variable. In aquacultural systems, the stock, which is misleadingly recreated, is done restricting. It no longer comprises the regular asset. Appropriated, it becomes assimilable to capital. In extensive development, it is the essential and auxiliary productions (microscopic fish and benthos) filling in as nourishment for the stocks which limit production (Llorente & Luna, 2016). In intense systems where the food is added, it is the wastes (unconsumed food, defecation, anti-infection agents and prescriptions, and so forth), just as departures of people who may alter the hereditary legacy of the wild populations, that should be acclimated to the acknowledgement limit of the fish biological system. All through the cycle of domestication, the capital and workforce of the activity increment simultaneously as the specialized escalation. The increment in usefulness (per unit of the region) in this manner got is joined by an expansion in the tension on the climate.

Figure 5.9: A rendering of the floating ponds vertical fish farm conceptualized by Surbana Jurong and Apollo Aquaculture Group. Their model of urban aquaculture has reported yields six times higher than traditional farming, using the same land area. Image: Surbana Jurong.

Source: https://www.eco-business.com/news/could-vertical-fish-farms-be-singapores-answer-to-food-security/

Throughout this interaction, the financial and social association of the activity likewise changes. As the specialized controls differentiate, the standards which characterize and support their activity - and specifically the systems administering the responsibility for regular asset - change. Experts along these lines depict the connections which have been created in horticulture between the training of the regular assets, the escalation of their misuse, and the land proprietorship system: 'It rises out of this long history that, since the Neolithic age, the "possession" of the land has bit by bit reached out to the various classes of land as they have become artificialised: based ashore above all else, then, at those point nurseries and walled in areas advanced and developed every year, cleared land bearing yields, created roughage knolls, land developed between two crude or neglected periods, land developed persistently, further developed fields, created and kept up with woods. The antiquated privileges of normal use (chasing, gathering, gathering, utilization of the normal, right to gather wood), as far as it matters for them, have consistently concerned land where, with no specific work, wood, regular grasses, the wild grasses of decrepit land, game, and so on kept on growing unexpectedly" (italicised by the creators). The historical backdrop of shellfish culture in Yerseke in the Netherlands shows that the cycles in aquaculture don't contrast on a very basic level from those saw in horticulture. Throughout the interaction of specialized heightening, the designs of the production, use and guideline systems were changed in stages:

- before 1860: aggregate use of regular beds by little neighborhood fishermen; guideline of access by the organization;
- from 1860 to 1933: with contrasts between clam culture and mussel culture, progress from fishing to extensive aquaculture, with the public closeout of individual concessions in the public oceanic field, and privatization of the stocks; application by shellfish culture cooperatives of individual production shares pointed toward controlling prices;
- 1933 - 1967: increase of production; epizootic illnesses; managerial guideline of production and markets;
- after 1967: vertical reconciliation of endeavors, which become free enterprise; a decrease of public mediation.

The portrayal of extensive aquaculture

Extensive aquaculture is described by a production line in which human mediation is focused on the proliferation of the stock. This may take two structures:

- through the actual formation of specific living srates: occasional shutting of tidal ponds to hold the adolescents before their relocation to the ocean (viticulture, which incorporates different intercessions like the guideline of the water flow, the disposal of unfortunate species, or the arrangement of extra food, and so forth), production of fry lakes and artificial nurseries to help the propagation of normal populations (salmon), catch of regular altercation by giving foundations to the assortment of hatchlings (shellfish culture), making of specialities for the adolescents of specific species when the density of the previous is a restricting component, and so on (Liu et al., 2015).

- by delivering incubation centre fry into the regular environment.

Extensive urban aquaculture

Extensive aquaculture is drilled in various urban settings; the most outstanding methodology comprises of loading fish in supplies and huge urban water bodies, trailed by recovery after a time of 1-2 years. Records of loading and reaping fish from urban supplies have come from urban communities like Brasilia, Brazil; Hanoi, Vietnam and Wuhan, China. Thinking about Wuhan, culture-based fisheries in Donghu Lake (1,500 ha) are reliant upon loading a great many silver carp (*Hypophthalmichthys molitrix*) and bighead carp (*Aristichthys nobilis*) seed and giving nursery regions in dammed bays, net-banished bayous and net confines to guarantee fingerlings are possibly delivered when they are adequately huge to stay away from predation (>13 cm). The control of predator fish is likewise attempted to assist with restricting mortality, while mass collecting is embraced following a year when fish are around 1 kg in weight. Attributable to improved administration, production expanded from 180 tones in 1971 to 1,840 tones in 1995.

The control of fish stocks in urban repositories, through particular stocking and gathering, has likewise been utilized to control eutrophication (Starling, 1998). A genuine imperative to aquaculture in urban supplies is the different utilization of such water bodies by different gatherings, regularly with clashing interests. The transparency of such systems likewise makes it hard for those cultivating fish to screen dangers, for example, conceivable contamination sources, or to watch out for the exercises of different clients. The utilization of enclosures or pens may comprise a chance for farmers to acquire selective admittance to parts of normal property assets, however, access of this sort might be hard to arrange and is probably going to cause clashes and conceivably hindrance less fortunate segments of networks. There is a developing collection of writing concerning normal property assets and rules and best administration rehearses proposed for sea-going assets might help create fair access and the board techniques for urban aquaculture. Notwithstanding, the consistent trimming of tilapias and other self-supporting fish stocks in eutrophic urban waterbodies is presumably perhaps the most useful, and valuable, framework got to by poor people in Asias' urban communities.

Confine culture is rehearsed for a huge scope in the Saguling-Cirata-Jatiluhur chain of supplies downstream of Bandung, Indonesia; estimates recommend somewhere in the range of 4,425 fish confines, delivering a sum of 6,000 t y-1 of tilapia (*Oreochromis* sp.), are available in the Saguling Reservoir. In any case, pens and pens are available to the more extensive climate and as such powerless to water quality issues. On account of Saguling, huge scope fish kills during the long stretches of January and February have been accounted for, albeit the specific reason has not been recognized (Lafont et al., 2019). This accepted privatization of the normal pool asset requires capital resources less accessible to poor people, who can hence be immediately barred. Even though past the extent of the momentum survey note here that urban wastewater, through supplement improvement of getting water-bodies, can upgrade production from wild fisheries. Nonetheless, little work has been done to measure the degree and meaning of this relationship, while most consideration is given to guaranteeing that supplement improvement in getting water-bodies doesn't surpass the ecological conveying limit, corrupting the climate and hurting catch fisheries.

Considering the far and wide absence of wastewater treatment offices in many agricultural nations, almost certainly, seepage water from urban communities in numerous nations is streaming to waterways, lakes, lakes and rice fields in urban and peri-urban regions, be that as it may, little work has been done to record or evaluate this, or to survey the expenses and advantages of casual wastewater reuse through aquaculture.

5.7 BENEFITS OF URBAN AQUACULTURE

Work, income generation and food security comprise significant advantages, specifically for individuals from less fortunate networks. Be that as it may, more extensive advantages stood to society, for example, overseen waste reuse, prompting further developed public and natural wellbeing security; financial advantages, for example, expanded assessment income and sponsored waste control; non-inexhaustible asset recuperation; extra utilitarian and non-useful qualities may likewise be credited to urban aquaculture. The World Bank has recognized the need to remember a more extensive scope of issues for financial choices and to amend the monetary examination of activities to incorporate externalities and maintainability. From the audit introduced here, it is obvious that a more careful appraisal of advantages related to urban aquaculture is expected to advise organizers and strategy producers regarding the genuine value of this action, both for needy individuals and society overall.

Figure 5.10: Urban Organics aquaponics farm using Pentair's advanced water filtration technology won the Agriculture Project of the Year in 2017.

Source: https://www.hatcheryinternational.com/pentair-sells-off-aquaculture-business-3513/

5.8 CONCLUSION

The investigation of broad systems makes it conceivable to evaluate the possibilities and distinguish and rank the conditions important for the turn of events and justification of this method of production.

The most conceivable theory we can advance based on the speculations on the guideline of the marine populations and the degree of production of development systems is that, from the biological stance, the capability of broad aquaculture could be tantamount with or shockingly better than that of fishing. While the broad development of new species leads to complex inquiries, in general, these issues can be settled by marine nature.

The specialized issues don't comprise an insuperable snag by the same token. A few exceptionally useful systems depend on basic procedures. Decent numerous R&D programs which have not brought about business production have been effective from the specialized point of view. The rundown of species whose propagation is dominated is extensively more than that of the economically developed species and is getting consistently more with the advancement of escalated aquaculture (Kaminski et al., 2018).

The hindrance is at this point not financial. Like shellfish culture, a few broad systems are entirely beneficial. The strategy is transposable to different types of extraordinary business value. The shortfall of ocean items and the exorbitant costs which are probably going to persevere in Europe for species that are not being delivered by intense development is great for the expansion of broad aquaculture. From this viewpoint, broad aquaculture presumably offers more adaptability than intense systems.

In any case, the real advancement of this method of production isn't following these ends. The most evolved systems are old and provincial. Ongoing exploration has not prompted any huge production in Europe. The lone marine development of vagile species which has arrived at a specific degree of improvement - salmon farming - likewise goes back quite a few years. For Europe in general, its production is presumably less by a significant degree than that of shellfish culture.

The principal impediments appear to be legitimate and institutional. An examination of the principle development systems uncovers the presence of solid connections between's the degree of advancement of the production cycle - and henceforth the interaction of training -, the financial and social construction of the activity, the improvement of individual possession and the simplicity of allocation of the stock. Along these lines, shellfish culture is portrayed by the combination of private farmers, proprietors of a stock whose appointment is worked with by its inactivity. Then again, where the issues emerge is in the change of the shellfish cultivating stocks to the biotic limit of the bowls and the portion of the inherent limits of the biological systems between shellfish cultivating and contending employments. These issues have not gotten the consideration they merit taking into account what is in question. The purposes behind this are to be found in the smoothness of the biological systems that entangles their appointment, and the fluffy idea of their legitimate status, which keeps the undeniable choices from being taken. The restocking of salmon, as far as it matters for its, is focused on the protection of the wild fauna and the help of stocks misused in fisheries where access is ineffectively controlled. The softness of the lawful status of the stocks, predation by outside fishermen, and the poor correspondence between the public expense of raising the fry and the private increases of the administrators block the financing of restocking. On the other hand, the better outcomes got by Japan in the advancement of the broad aquaculture are associated with the aggregate possession system for natural resources that this nation has received.

When, during the 60s and 70s, the tumble off in halieutic production brought about another excitement for aquaculture, the correlation with horticulture was frequently summoned to legitimize an increase of the R&D exertion to assist marine aquaculture. While the examination of the systems of production does surely affirm the similitude of the cycle of increase in the marine and land systems, it additionally uncovers contrasts that discredit the relationship with the Neolithic Revolution.

The principal distinction is that the cycle of artificialisation began a lot later in the ocean than ashore because until after the Second World War, the interest in ocean items could be fulfilled by the expansion of fishing activities. Up to that time, social orders had accordingly had the option to manage without domestication - aside from shellfish culture whose improvement was basic. Today, misuse of the living resources of the sea - regardless of whether it involves the justification of developing systems like fishing or shellfish culture or the domestication of new species -, the protection of the climate and the designation of its restricted limits between continually developing and expanding utilizes require the selection of establishments that make it conceivable to viably control access.

The subsequent contrast is that the natural resourcesof the ocean are on a lot more notevaluey scale than earthly assets. The ease of the environment and the portability of the stocks entangle domestication and the improvement of suitable foundations. The arrangement embraced a few centuries prior in horticulture, for example, the sharing of the land is not any more transposable to aquaculture - except for the responsibility for stocks than to fishing. However, there are alternate methods of accomplishing similar finishes.

The third distinction is that new examinations on the starting points of horticulture and the development of the first creating social orders show the restrictions of clarifications of the Neolithic Revolution dependent on absolutely (specialized advancement) or monetary (reaction to natural, organic or segment pressures) contemplations The position of the improvement of establishments administering earthbound exercises (according to the prior advancement of farming, handed-off by that of exchange and industry) most likely clarifies why the principal public approaches for the advancement of marine aquaculture had the option to consider as appropriate for the ocean organizations which had filled in as the establishment for the advancement of the earthly exercises, however, which didn't exist at sea.

At the worldwide level, the expansion of public legislation comprises a noteworthy stage in the misuse of the seas. It gives oceanic expresses the chance of presenting in their regional waters the systems of eliteness essential for the advancement of domestication and the defence of their employment. In any case, acclimation to the new conditions is as yet in the beginning phases. With the systems in power, the private and social expenses and advantages concur gravely: would-be fish farmers can't be given the ensures which would instigate them to put resources into new systems, while in the development systems producers have an interest in expanding their stocks past the base vital for agreeably abusing the trophic limit of the environment. The challenges experienced by open experts in playing out their capacity as stewards of the regular legacy, and the tendency of the organizations to incline toward set up utilizations to the disservice of seriously encouraging new productions are clarified by the softness of the systems of selectiveness and the inadequacy of managerial guideline as a strategy for distributing privileges of utilization (Jothiswaran et al., 2020).

With regards to changing the utilizations to the limit of the normal assets, the viability of the guideline structures relies upon the fortuitous event between their srate of ability and the spatial construction of the regular assets. Given the conspicuous variety of the sizes of the unitary assets and the following of political outskirts, the presence of such a correspondence involves possibility. Truly, two units rule the design of the assets - the animal or vegetable population, and the environment; these units are free of the utilizations and their scales make it conceivable to visualize a significant decentralization of their turn of events - at the nearby level for seaside assets, and territorial (common) for the assets of the mainland rack.

The production of free open offices, constrained by specialists comprised of delegates of the equipped organizations and client gatherings to be liable for dispensing privileges of utilization, would be an answer for these issues. This game plan is ordinarily utilized in the misuse and preservation of earthbound assets, for example, woodlands or water assets. The combination of improvement would be advanced if the abilities of these offices covered natural resourcesautonomously of their employments. As stewards of the regular legacy, they could, through extricating the normal shortage lease of the assets, viably assault the main driver of the overexploitation and utilize this income to cover the consumptions intrinsic in their capacities (observing and assessment of assets and their uses, designation of rights, control and observation), and money programs for the misuse of natural resources(restocking and overpopulation, annihilation of non-native species).

It is improbable anyway that the positive potential for broad aquaculture will get the job done to trigger such changes. Be that as it may, assessment of the lost open doors in broad aquaculture may assist with advancing raise mindfulness previously stimulated by the conspicuous insufficiency of the current control of fishing or the preservation of the marine climate.

Thinking on institutional changes is as of now all around cutting edge. A few nations have attempted to change their fisheries guidelines. Considering the similitude and complementarity of the issues, the investigations would acquire if the guideline of the aquacultural systems and the preservation of the climate were important for similar projects. The fundamental hypothetical information and the important field observations are accessible. A near investigation of the foundations covering the entirety of the employments of the normal assets could prompt a logical network which would make it conceivable to analyze the benefits of the various strategies for a guideline that might be imagined as an element of predefined rules and to comprehend the states of development of the establishments in various settings. The compensation of this information would fuel the discussion with the gatherings concerned (nearby specialists, organizations and client gatherings).

The time of change is probably going to be long nonetheless. This time could be effectively utilized for better agreement and talking about the issues of the turn of events and association of broad aquaculture. The execution of painstakingly select programs, covering all parts of the issue, could help here. Some could be worried about the expansion of shellfish culture - to exploit the simplicity of assignment of the stock -, others with the aggregate use of vagile species - to assess the capability of this framework. In this last case, it is desirable over pick species whose srate of migration is coast and confined. In the two cases, types of high business value or liable to bring about exercises for which there is a feasible interest (sport fishing) ought to be given need. The scallop would be a decent possibility for the principal program, and the turbot would almost certainly be a decent decision for the second. Plausibility studies ought to go past the specialized and monetary angles and examine the lucidness between the production evenhanded, the srate of conveyance of the stock, the possession system of the regular asset, the financial and social design of the activity, and the component for assigning of the privileges of utilization. These projects would be pointed toward showing the natural and specialized achievability and the financial interest of aquaculture, just as assessing the monetary results of the insufficiencies of the guideline systems now in power (Jensen et al., 2014). Thus, have great correspondence between the groups concerned - including the overall population. As prior programs have shown, incorporated investigations of this sort may propel information on inquiries of incredible logical and monetary premium, just as the guideline systems for marine populations, the association of fishing, the training of workers, the battle against the colonization of extraordinary species, and the protection of the climate. However reassuring discussion on issues like the job of possession during the time spent specialized heightening and the protection of the climate could add to the progression of deduction on delicate issues, for example, monetary improvement measures. The achievement of such projects will rely upon their cognizance, the thoroughness of their origination, and the activation and coordination of the disciplines fundamental for their execution. For this, it gives off an impression of being fundamental for use systems examination strategies and a multidisciplinary approach.

Chapter 6

The Cost of Practising Extensive Aquaculture

CONTENTS

There are several costs incurred by farmers practicing extensive aquaculture. Most of the costs go into maintaining the system. There are various sectors where the farmer will have to spend large sums of money. At the start of the production system, the farmer will need a large capital to purchase most of the tools needed such as cages. However, with the increase in awareness of the need and effective methods of aquaculture production in most parts of the world, farmers can be more aware of where they will be spending their money. There are also news highlights on the increase in fuel costs needed in running the aquaculture production systems. In extensive aquaculture, most of the focus is placed on fish reproducing and them being sold in markets. If not done properly, farmers could be incurring losses from these farming systems. In most cases, farmers are advised to hire financial managers who will help them handle issues pertaining to the purchase of farm tools.

Figure 6.1: Various activities done in extensive aquatic farms require routine maintenance.

Source: https://www.shutterstock.com/search/extensive+aquaculture

Depending on whether the farming system has been done on either small-scale or large-scale, the cost of the farming system will vary. Economic stability by the farmer to maintain the farm system is needed for the farming system to actualize its full potential. If the farm is to attain full profits then the manner in which funds are used should be a major concern. Cost production is known to affect farmers' profits. For a farmer to achieve good cost management they need to look into certain factors that will determine how much money they will spend to maintain the farm. These factors are usually covered under site selection (Holmen et al., 2018).

6.1 ECONOMIC CONSIDER IN SITE SELECTION

For a farmer to minimize future costs on extensive aquaculture production, some factors considered under site selection are required. These factors include soil conditions, hydrology and topography. These conditions are considered to be bio-physical parameters. These factors will determine whether the farmer will have to spend more money on maintaining the farm.

6.1.1 Topography

Topography is an essential element considered by farmers when setting up farms. Usually, they are advised to establish these farms in areas where there is high gravitational influence. The gravitational flow is useful in allowing water to flow into and leave fish ponds. This saves farmers the cost of purchasing pumps and other systems to allow water to flow into ponds. They are usually encouraged to establish fish ponds at an elevation level higher than that of the water table. This will enable it to be properly drained into ponds without needing external forces. When looking at the topography, the farmer checks whether it is hilly, undulating, sloping and flat. It is economical for farmers to choose an area where the ground is flat or slightly sloping. If the area is sloping, the slope should be around 0.5 to 1.0%. If the slope is more than 2%, the farmer will have to incur more costs during the construction and maintenance of the pond.

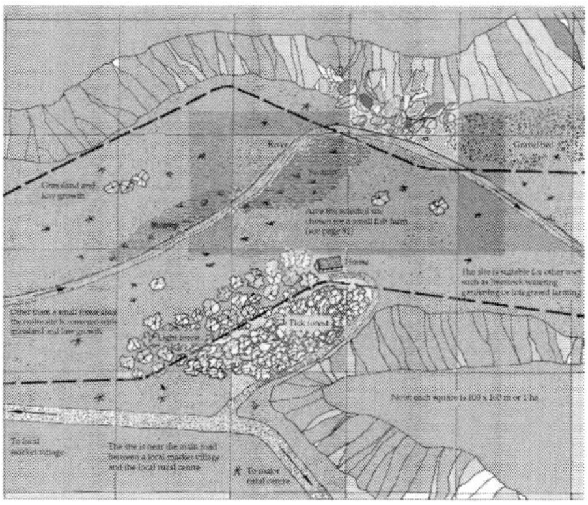

Figure 6.2: Good topography is needed in reducing maintenance and repair costs for farmers.

Source: http://www.fao.org/fishery/docs/CDrom/FAO_Training/FAO_Training/General/x6708e/x6708e01.htm

For a farmer considering marine farming then it would be advisable if the pond is established in a flat area close to water bodies. Ideal sloping is needed for farmers to incur lower costs. Highly sloping areas should be avoided by farmers as they could incur high costs during the construction process. This is because they are required to cut vegetation and fill soil in some areas. Also, areas having high vegetation should be highly avoided. This includes areas with big trees as well as those with thick vegetation. High costs will be incurred in clearing those sites. The process is also quite difficult. Further, areas chosen for farming should have good elevation. This will lower the cost of making frequent repairs when flooding or high rainfall is experienced. Extensive aquatic farms established in poor elevation grounds will require farmers to frequently construct and repair dikes to prevent flooding during high tides (Hall et al., 2021). The same applies to areas with very small tidal functions as the farmer will have to spend extra costs in draining and filling the pond. In most cases, the draining cannot be fully executed. For this reason, farmers are encouraged to set up farms in areas where tidal functions are moderate. Moderate tidal fluctuations allow proper management of such ponds at a lower cost. For extensive fish farming systems the ideal amplitude of 1.5 meters.

Therefore farmers should avoid areas of high flooding probabilities. Other than increasing costs of draining and filling ponds, they could also affect water quality, the introduction of undesirable fish species and also erosion possibilities. Good engineering skills should be used in the process allowing proper draining of ponds. Farmers who establish their farms in non-ideal areas, they will have to incur extra costs due to the pumping of water from higher areas to lower areas and vice versa. The same applies if the farmer sets up the pond in areas lacking gravitational flow. When using pumps, the farmer will have to incur the cost of replacing broken parts of the pump, purchasing new pumps, using electricity while pumping water and also the cost of hiring an individual to install, fix and check for the proper running of the tank. Depending on the number of pumps used in the pond, the farmer will have to incur more costs if there are several pumps.

6.1.2 Hydrology

With fish farms, water is an essential element needed in ensuring the survival of fish species. As an essential commodity, it should be readily available for farmers. This is why most fish farmers opt to establish their farms close to water bodies. It saves them the cost of having to transport water from water sources to the farms. However, farmers should take into consideration the type of fish species kept. Access to constant water supply is also key in helping farmers save costs. There are cases where farmers have to purchase water as their supply is limited. As large gallons of water are needed in ponds, purchasing water means that they will have to incur relatively high costs. For farmers who have land close to water sources or reservoirs, they will be able to easily handle issues pertaining to water such as water loss through evaporation, drainage and seepage. They also need to have a backup plan as surface water may not be available throughout. There are constant fluctuations depending on the season. There may also be changes in quality.

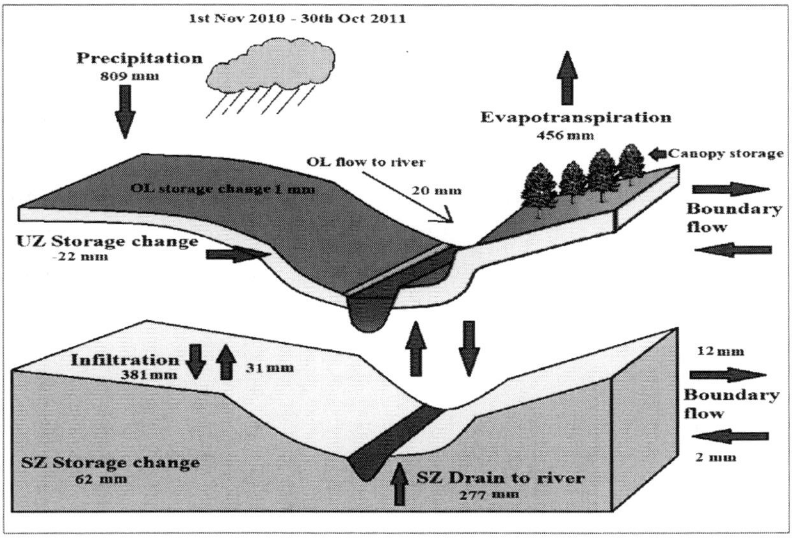

Figure 6.3: Establishment of wells helps ensure continuous water supply.

Source: https://www.mdpi.com/2076-3417/10/4/1281/htm

To ensure continuous water supply, most farmers have established well to help them ensure continuous water supply. They, therefore, have to incur the cost of constructing a well. This includes purchasing materials as well as the labor involved in the construction process. This water source is highly preferred as it is said to be free from diseases, pollutants, predators and parasites among others. However, when using wells, they will have to purchase pumps to move water to the ponds. High expenses will be incurred by the farmer when drilling the wells. Maintenance costs are not usually high if the well and pumps are properly used. It is also advisable if the farmers set up a farm in areas where there is less effluent discharge. This is because if these effluents find their way into the ponds there is a high likelihood that it could result in fish suffering from various diseases (Gjedrem, 2012). It will require farmers to incur high costs when treating water sources. Water treatments are usually expensive therefore avoiding these effluents could really save the farmers from very high costs. Farmers in developed countries have been smart in water usage in that they use deep-well injection of effluents. Though this practice increase the initial capital costs as well as that of pumping, it ensures the aquaculture system runs smoothly.

Most small-scale farmers may not be able to construct wells. For this reason, they opt to establish storage tanks. Though they may be expensive to purchase, in the long run, they save the farmer from extra costs such as purchasing water. If possible, the farmer should establish their farm in an area that experiences low evaporation as it will help reduce water loss. Areas known to experience high rainfall are highly recommended as less effort is required to maintain the water level. For farmers using pumps, other than costs on operating the pumps they also need to follow up on fuel prices as the pumps use fuel to run. An increase in fuel prices affects farmers as they will have to pay more to get fuel. Fuel costs are usually calculated under pumping costs. Fuel costs usually depend on the type of fuel used. Using diesel fuels will be quite expensive to farmers as they are usually costly. There are pumps that run on electricity. This is a good alternative for farmers though high prices of electricity will affect them. The best option would be to use solar-powered pumps as they are more economical.

Figure 6.4: Solar-powered pumps are cheap energy alternatives for farmers.

Source: https://blogs.worldbank.org/water/solar-pumping-101-what-why-and-how

The advantage of using solar-powered pumps is that solar energy can be freely accessed by all. The capital to purchase the solar panels will be high. The maintenance cost is very low. The other advantage of solar-powered energy sources is the fact that they run for many years. There are also batteries that have been developed allowing energy storage more so in times when there are low sunshine amounts.

6.1.3 Soil Conditions

Another vital element considered by farmers when establishing extensive fish farms is soil conditions. In the case of fish farms, the farmer needs to consider water retention and fertility. The farmer needs to consider the benthic community useful in ensuring the survival of most fish species. This community helps supplement diets. They are usually a source of food and if present in large amounts could help farmers supplement fish diets. This goes a long way in reducing the amount of money that could be spent in purchasing fish feeds. Fertility is usually determined by soil composition. Highly organic soils are known to contain high amounts of nutrients that could slip into water resulting in the growth of aquatic plants. However, it is not advisable for farmers to use highly organic soils in the construction of fish ponds because these nutrients usually undergo processes such as decomposition which could lead to ponds cracking or exhibiting high levels of seepage. This increases costs that could be incurred by farmers in maintaining the pond. They will have to cover the cracks frequently. With regards to seepage, it means that the farmer will have to incur high costs in filling and draining the pond.

Figure 6.5: Clay soil is mostly used in the construction of ponds because of its properties.

Source: http://agropedia.iitk.ac.in/content/fish-pond-construction

It is more advisable for a farmer to establish their pond in an area having clay soil as they have high retention abilities. As it allows little water to pass through it, it is more convenient for small-scale farmers as they will spend less on purchasing materials such as black paper. Black heavy polythene papers are usually purchased by farmers to be used in soils having low water retention abilities. The paper is usually checked to ascertain that there will be no leakage. In large-scale farms, farmers usually go to the extent of constructing ponds made of cement. High levels of certainty and keenness are required when constructing these ponds as poor construction could result in farmers losing large sums of money in fixing the ponds. Ponds constructed purely using clays soils are mostly small-scale farms as refined establishments are usually very expensive. Other factors considered during site selection include the cost of the land. The farmer should consider how much they would be paying for the land they intend to use. The price should be relative to the size and its location. Farmers are usually advised to avoid pieces of land having competitive uses (Gopakumar, 2003).

Among the major factors affecting the cost of production when constructing the pond include the depth, shape and size of the pond. Ranging to the large scale of small scale, the cost of fish ponds varies. Usually, the larger the size of the fish pond the higher the amount of money going into construction. This is because more materials are used during the construction process. Also, large amounts of water are used during the construction process. The size of land used in large-scale farms will also cost more than an ordinary pond. Small-scale farms can also be quite costly depending on the type of pond being constructed and the type of materials used. The advantage of large ponds is that less land is used in creating ditches. Compared to small-scale farms, it is more convenient and less costly for the farmer in the aspect of equipment use and purchase.

When bought in bulk, these equipment will be much cheaper than purchasing a single one. This means that there is a depreciation in the amount of capital investments. Though is much cheaper for a farmer to run a large-scale farm compared to a small-scale one, there is a significant difference in the output in that there is a high likelihood of output being lower due to some mistakes that may be made through the course of the production. Some of these errors include improper draining of ponds. This may be due to the fact that ponds usually take a long time for them to be drained. In other cases, poor results are due to the unsatisfactory ratio of the shoreline. This means that the shoreline of a big pond is usually shorter when compared to those from several small ponds. High levels of productivity are usually experienced along the shorelines as there is a higher concentration of food along the lines. Also with large ponds, it is quite difficult to ensure even distribution of food. For this reason, farmers managing large ponds are more likely to incur losses compared to those from small-scale farms. They have to incur higher maintenance costs compared to those having small-scale farms.

The major issue with small-scale farms is that the initial construction is usually high. This is because small ponds need more land area. It is also quite difficult for the farmer to introduce various technologies useful in improving farming operations. The cost of construction may increase or decrease depending on the climatic conditions of the area and also the various species of fish reared in the pond. If the pond is to be constructed in a cold area then the cost will be lower as a shallow pond will be constructed. However, in warmer areas deeper ponds are constructed. Such ponds are usually costly. There are cases where prices vary depending on the shape of the pond.

6.2 FEEDING COSTS

Figure 6.6: For quality fish to be harvested, fish need to be well fed.

Source: https://www.fishlaboratory.com/fish/types-of-fish-feed/

To ensure the survival and wellbeing of fish, the farmer has to ensure that they are all well fed. Usually, feeds are bought depending on the species of fish. Although extensive aquaculture fish production systems are low input compared to intensive aquaculture systems, a large amount of money usually goes to purchasing feeds. As aquaculture productions systems have become dependent on feeds given to fish, increase or decrease in fish prices could severely affect production. Farms less affected by changes in feed prices are those that are not dependent on feed prices. With the growing increase in fisheries, there has been a growing increase in the need for fish feeds. In most cases, aquatic plants are not considered as aquatic feeds bought by farmers as they are naturally occurring. They are usually not depended on as a sufficient source of food as they may be in insufficient amounts hence they cannot sustain large fish populations.

In ponds where they are present, the farmer purchases feed to help sustain the already existing food supply. There are certain countries that use naturally occurring plants though they are produced on a large scale and sold to other countries.

China is among the countries known to be a producer of aquatic fish products. China is said to be a dominant producer of aquatic plant products. With the increasing demand for aquatic feeds, there has been an increase in the prices of aquatic feeds.

The increase in prices is a result of the high cost of ingredients used in making aquatic feeds. High prices of such commodities are attributed to the global increase in prices of energy and commodities. According to some data, the increase in commodities is by 50% (Frankic & Hershner, 2003).

There are various fish products sold in local markets as well as the international market. Some of these products include dry foods such as flake foods, produced globally. In most cases, the flakes are baked so that there is little moisture in them.

They are usually sold in different forms among them wafers, granules and pellets. There are vacation foods also referred to as food blocks. Such feeds are not commonly used as they could result in water pollution when left in water for longer durations of time. For perishable fish feeds, they are made and sold as frozen commodities. In most cases, perishable fish feeds are made of ingredients such as brine shrimp, Daphnia and bloodworms. Among the ingredients found in fish feeds are nutrients such as carbohydrates, fats and amino acids. A keen analysis of fish farmers' production costs could show that fish feeds usually account for 50-70% of the costs. There has been a noted jump in the increase in prices of fish feeds.

Figure 6.7: Fish feed prices vary depending on the type of commodities used in making them.

Source: https://www.fish-feed-extruder.com/Application/self-made-fish-feed.html

From recent statistics, it is evident that the aquaculture sector is independent of various global phenomena. The major issue has become the manner in which global tendencies will affect the aquaculture sector. Hopes of a decline in prices are not expected anytime soon. This is due to the constant increase in the prices of commodities. There is a variation, in the prices of fish feeds. Prices are usually dependent on the type of fish species being fed. The aquaculture sector is usually dependent on common commodities including wheat, rice, fish, and fishmeal, corn and soya beans. Prices of these commodities have increased due to various concomitant factors. The increase in the price of these commodities has been brought about by sudden weather changes and reduced harvesting of commodities. The growing increase in household income in developed countries more so in China has helped improve the demand for staple feed stuffs. An increase in prices is anticipated to be reflected in global markets.

The increase in prices in other commodities such as fuels has contributed to the increase in fish feed prices. America, Brazil, United States, Chile and Argentina are known to be major producers of ingredients such as soybeans and corn. In the case of Brazil, commodities are exported to other countries. Brazil has a port are they make heavy use of trucks in transporting soybeans to various areas. Therefore, the farmers have to deal with the cost of transporting farm produce. They therefore would incur the cost of changes in fuel prices. In some countries, this may not be the case as farmers sell the produce to the government who will then set a new price for the commodities before they are sold in other countries. The projected increase in global demand for fish and fish products has also led to the increased price of fish commodities. From the year 2000-2005, the price of fish feeds in the United States ranged from 500-700 USD for a ton of these products. Over the years the prices have increased so much that they have doubled (Fore et al., 2018).

Fish prices usually vary depending on the quality of the feeds. High-quality feeds are relatively expensive as they have more ingredients. Also, a lot of work goes into refining these products so that the nutrients balance meets the demand. In as much as farmers have to bear the high cost of the products, manufactures also have to incur a lot of challenges. They have to strategically plan the production process so that they are able to sell the products at a suitable price. This means that they have to optimally use ingredients so that they can get the maximum amount of products out of little amounts of ingredients. To some great extent, fish feed prices have been experiencing a growing trajectory due to the increase in prices of ingredients like fish oil. Manufacturers also have to bear the cost of packaging and marketing fish products for farmers in an attractive manner. If they poorly set prices of commodities, they will have to incur losses. They also have to deal with the issue of the high cost of these products in that there is a high likelihood of farmers not purchasing certain feeds.

In some cases, small scale-farmers may not be able to incur the high cost of these fish products. There are those who opt to purchase the ingredients themselves and make the products themselves. This is a more economical method of creating fish feeds. When making these feeds, farmers consider the species of fish kept. There are those who rely on aquatic plant products. The fish products are made to supplement the products. In this case, farmers make use of natural aquatic products. The advantage of these feeds is that they are highly nutritious as they cater to fish nutrient demands. In as much as the products are self-made, they are usually made in different in that they can be granular feeds, biscuit-like, dough–like or powder. Those in powder form are made by mixing ingredients and grounding them together and adding a certain amount of water before they are given to fish. This form of feed is usually to fish feeding plankton, little fish and fish fry. Other types of powdered feeds can be given to eels, carnivorous fish and shrimp.

Granular feeds are mostly made of fish pellets. The farmer purchases the various fish pellets and mix them according to fish requirements. Depending on the types of pellets mixed, there can either be hard or soft pellet feeds. There expanded and micro pellet feeds. In as much as they are self-made feeds, the farmer needs to ensure that they meet the nutritional needs of the reared fish. If there are various fish species reared, the farmer has to take into consideration the variation and incorporate ingredients accordingly. They also need to ensure that feeds are mixed at a suitable ratio. They should also consider the digestibility and palatability of fish. The formulation should also be balanced in that both amino acids ad proteins are balanced. Some framers have been able to make extra cash from making self-made fish feeds. They make large amounts of self-made fish feeds and sell them to other farmers. If done properly, this could help supplement the farmer's income.

The challenge with fish feeds is the growing increase in prices. This creates a problem for farmers as they may not be able to afford some of these products. Also if feeding is not properly done such that the feeds are improperly administered to fish, the farmer will incur losses. Underfeeding of fish results in fish exhibiting low development rates and also the death of the fish. Excessive feeding results in excessive amounts of feeds being used which will be very costly for farmers. It also causes eutrophication which limits the amount of light entering water bodies. The same applies to oxygen levels. This could result in the death of fish species and a reduced amount of aquatic plants in water bodies. Proper budgeting is needed in managing fish feeds.

6.3 COST OF EQUIPMENT

In most large-scale farms, various are used to make activities more productive and effective. Some of the equipment used in farms include transportation tanks, nets, portable and stationary pumps, freezers, ice machines, oxygen meters, pH meters and feeders among others. The type of equipment used in farms varies depending on the farmer's preference. When starting an extensive fish farm, the farmer needs to know that they will have to meet large costs on purchasing and maintaining farm equipment. For instance, one of the tools commonly used in fish farms is pumps. Pumps are very essential as they are useful in moving water from water sources to the ponds. The type of pump bought usually varies depending on the amount of water that should be pumped. In most cases, the pump comes with a motor that helps to drive water to various ponds (Fiorella et al., 2021).

Water circulation is achieved by using pumps. They are also useful in maintaining good oxygen levels in farms and reducing the chances of water stagnation. Water circulation is key in ensuring fish are very healthy. They are useful in the removal of contaminated water from ponds which exchanging it with fresh water. While selecting the type of pump to be used in the farm, there are certain elements taken into consideration. They need to ensure that the pump bought is properly functioning. The first factor taken into consideration is the type of pump. Depending on the manner in which the pump is to be connected, the farmer will decide on the type of pump to be bought. In most cases, external pumps are bought as they possess a higher capacity though they are very noisy. In rare cases submersible pumps are used. They are quite cheap and are more quiet in operation.

Figure 6.8: Different varieties of pumps can be used in directing water to fish ponds.

Source: https://airwaterfish.com/aerators-aquaculture/pumps/

Another factor to be taken into consideration is the flow of water. This is the amount of water to be driven into a pond. The farmer needs to take into consideration the total capacity of the fish pond. By so doing they can be able to establish the amount of water that can be held in the pond. The other factor to be taken into consideration is the head of the pump. This refers to the maximum distance the pump can push water. It is dependent on the height at which water has to be moved. Fish loads also need to be considered. This is the amount of fish kept per gallon of water. If a large number of fish is kept then there is a need for more water circulation. High flow rates are required in a large number of fish is kept. The need for filters can also determine the type of pump to be installed. If a higher flow rate is needed then a filter is required. The amount of power used in the farm will also determine the type of pump. After taking all these factors into consideration, the farmer will have to consider various brands and prices of the commodities.

Usually pump prices range from 30-200 dollars depending on the operating conditions, flow rate and capacity. Pumps produced by companies such Aqua space, Pond Master Pumps and Alpine pumps are specially catered for the aquaculture sector. Companies such as All Pond Solutions

Inc. are well-known retail pump providers. They make various products used in ponds such as tanks, skimmers and filters. Pumps produced by this company have a flow rate of about 2000-16000GPH. Pumps produced by this company are of high capacity and are used in large-scale farms. Pumps produced by Alpine pumps are usually oil-free and use magnets in driving the pumps. These pumps have filters fitted in them to avoid clogging of water. Purchasing of pumps is relatively expensive initially. The farmer has to deal with maintenance costs. These costs include replacement of propellers or fixing of damaged parts. If the pump has filters, they have to conduct routine cleaning of the filters. By using pumps, farmers have to deal with electricity costs. In this case, it may be more convenient if they use solar-powered pumps as it would save them the cost of having to pay electricity bills (Erondu & Anyanwu, 2005).

Farmers also have to incur the cost of purchasing and maintaining fishnets. Prices of fishnets usually vary depending on the size and quality of the net. There are various types of nets used in the aquaculture sector namely: braided dyneema netting, twisted dyneema netting, braided nylon netting and twisted nylon netting. High costs are incurred when a farmer purchases a net that is thick in nature. This is because these nets are made of high-density materials that is expensive to purchase. A decline in cost is experienced when thin nets are used. Thin nets also need fewer anti-fouling chemicals. The heavier and thicker the net, the more expensive the net will be. When purchasing nets, the farmer needs to consider the species of fish kept. There are those that can grow to huge sizes and are very heavy. In this case, the farmer will have to purchase a thick net. This is very essential in helping farmers avoid the need to purchase net frequently.

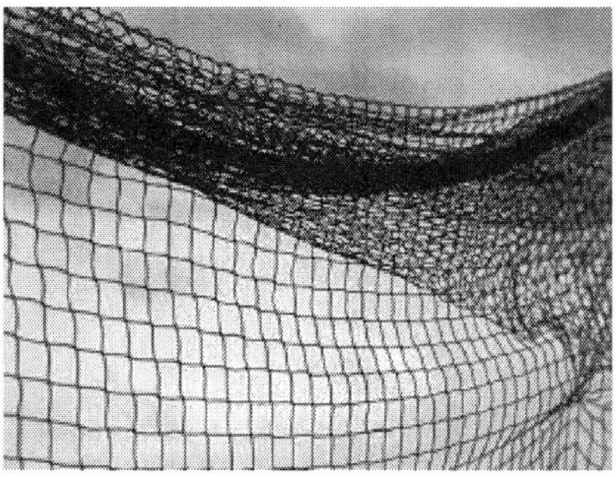

Figure 6.9: Nets are extensively used in fish harvesting.

Source: https://www.plastic-netting.org/plastic-mesh/fishing-net.html

Routine maintenance of nets involves farmers frequently washing them. These help to reduce breakage. If nets get torn, they need to be sown. In highly mechanized farms, other methods of harvesting are used. Among the qualities considered by farmers when purchasing nets are their strength, stable structure, softness, good tenacity, ease in operation, durability and good breaking length. Also, the farmer needs to look at the manner in which the nets have been twined and also the size of the mesh. This is very essential more so with regards to the manner of harvesting. Currently, farmers have been advised to pick a net of reasonable size so that little fish will not be killed in the harvesting process.

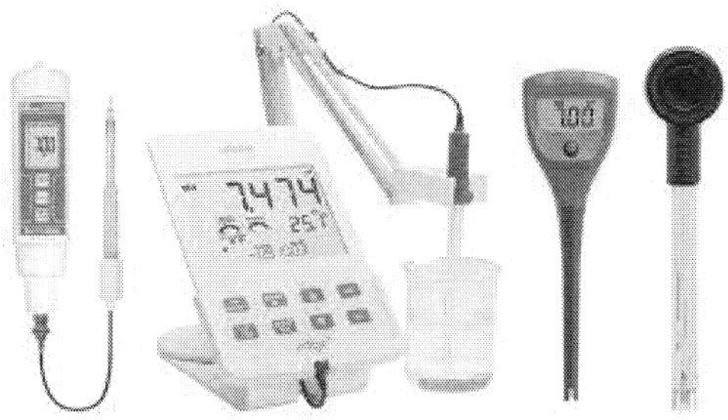

Figure 6.10: Water quality is essential in ensuring the survival of fish species.

Source: https://www.nhbs.com/blog/ph-meter-calibration-maintenance

Farmers also have to incur the cost of purchasing a pH meter. PH meters are heavily used in ensuring that water is of the required pH. This is very essential in ensuring the survival of fish species. Various methods can be used in measuring the level of acidity or alkalinity of water. However, it is quite difficult to accurately measure the pH of water. Initially, litmus paper was used in measuring the pH. The challenge of using this method is the fact that it cannot extremely accurate results. The paper is dipped in water and depending on the color change, one can note the pH of the water sample. The paper is made of a chemical-dependent on the pH of water. Another method that could be used is called titration. The challenge of using this method is the fact that it is very time-consuming and expensive in the long run. It also requires an individual with knowledge and experience to conduct the titration process. It is also dependent on the use of certain equipment to conduct the titration process.

For this reason, it was agreed that the most advisable method of measuring the pH of water is using a pH meter. This tool was developed while using the principle of an oxygen meter. The instrument is made of a probe having two electrodes. One of the electrodes is used in measuring the concentration of hydrogen ions while the other electrode is used in measuring the concentration of hydroxyl ions. Usually, there is a current that flows through the electrodes. The amount of current following is usually dependent on the concentration of hydrogen ions. The voltage across the electrodes is used in calculating the pH of the water sample. Usually, these tools are calibrated before they are used. The advantage of using this tool is that the farmer will purchase it once and does not have to deal with issues regarding pH meters. They are easy to maintain. Currently produced pH meters are portable and are easily available for farmers. They need frequent cleaning for results to meet the expected (Evers, 2010).

An oxygen meter is another tool needed in fish farms. This tool is used in measuring the amount of oxygen in water samples. These tools give vital information as oxygen levels in water determine the likely hood of fish surviving. They also help farmers known whether water samples need to be replaced or other forms of oxygen additions need to be involved. The kit is relatively expensive but there minimal costs when it comes to maintaining the technology. The advantage of using this tool is the fact that they run on batteries. There is no need for the farmer to incur costs on electricity usage though if it were to run on electricity, it would not take up large electricity units. An increase in the variety of such tools has helped lower the cost of such tools.

Figure 6.11: Storage tanks play a vital role in the transportation of fish products.

Source: https://xhyfrp.en.made-in-china.com/product/lKOEwxbuCkpP/China-Fish-Storage-Containers-FRP-Fish-Farming-Tank-Water-Tank-Fiberglass.html

Storage tanks play a number of roles. The major task includes them being used in transporting fish products from the farms to market areas. In other countries, rather than constructing ponds, farmers used storage tanks in rearing fish. They are also used in transporting fingerlings to new ponds or when the farmer is relocating the farm to a new area. Using storage tanks is an added cost to farmers as they are very expensive. Prices of storage tanks usually vary depending on the size in that large tanks cost more compared to small tanks. Another factor contributing to the high cost of these tanks is quality. High-quality storage tanks are those made using durable materials. This contributes to the high cost of these tanks. The type of material used in making the storage tank will also contribute to its price. Species of fish reared will also determine the type of tanks to be used. Fish species such as tilapia require relatively large storage tanks while others such as shrimps do not require large storage spaces.

There are storage tanks whose materials have been made so that they allow oxygen penetration into the tank. This helps to maintain high oxygen levels in such tanks. However, farmers do not have to purchase extremely expensive storage tanks. In Kenya, small-scale farmers have used plastic tanks in rearing fish. They purchase ordinary tanks at a lower price and modify them to meet their needs. This saves them from extra costs. In other areas, farmers use storage tanks made of stone. This form of farming is very expensive at the onset as the farmer will incur high costs in purchasing various equipment to be used during the construction process. It also requires a very skilled individual to be involved in the process of making the tanks. Poor construction could lead the farmer to debt as they will have to constantly repair damaged areas. In this case, it may be more convenient for one to purchase a storage tank.

Prices of storage tanks also vary depending on the capacity of water to be carried. Tanks that could carry large gallons of water are more expensive compared to those carrying less gallons of water. The farmer will have to decide on the type of tank they would like to purchase once they have decided on the type of tank they would like to purchase depending on the price. Also, the durability of the tank is important. Long-lasting tanks are more economical compared to tanks that are not durable. The farmer may have to incur costs on maintaining the tanks. In the case where there is a breakage, the ability of the farmer to fix the tank. In some cases, the tanks cannot be prepared the farmer will have to replace the tank. With storage tanks, there are those used in transporting fish to markets. As fish are perishable, the farmer needs to take up other methods when fish has to be transported over a long distance. In this case, refrigerated tanks are used (Ellingsen et al., 2009).

The farmer may not need to purchase the tanks as some transport trucks have already been fitted with refrigerated storage tanks. In the case where these trucks do not have refrigerated storage containers then the farmer will have to purchase the tanks themselves. These tanks are very expensive. They are also very costly to maintain and require a large power supply to run them. The maintenance cost of these storage tanks is very high as finding spare parts of these tanks is very hard. As they require large amounts of electricity, the farmer will have to incur a higher cost on electric bills. As mentioned earlier, it is very crucial for farmers to establish ponds in areas where there is managed release of effluents. Improper release of effluents will result in water contamination. Contamination could also be brought about by excessive feeding of fish resulting in excessive amounts of feeds present in the water supply. As fish are able to survive in certain conditions, it is very important for the farmer to routinely conduct water treatment.

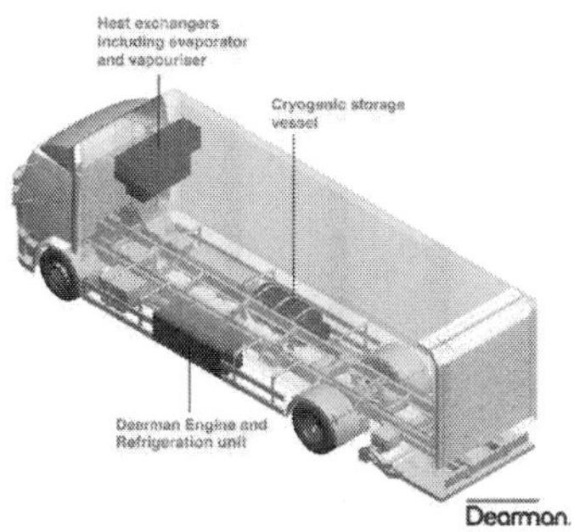

Figure 6.12: Refrigerated storage tanks are used in transporting fish over long distances.

Source: https://www.greencarcongress.com/2016/02/20160226-dearman.html

The challenge however is the fact that water treatment processes are relatively expensive. Tools used during the process are very expensive. Also, individuals involved during the treatment process have to possess water treatment skills. If water is to be transported from a treatment center to the pond then it will be relatively cheaper for the farmer as they would not have to incur the cost of purchasing tools used in the cleaning process. Also, at the treatment station, purified water will be sold at a reasonable price. However, if the farmer decides to conduct the purification and treatment process then costs will be relatively high. Chemicals used in the treatment process are very expensive. If the purification and treatment process is to be done at a certain frequency then the farming process will be relatively expensive.

They also have to cater to the cost of labor. Labor is highly required in extensive fish farms. There are various activities that require human labor. The amount of labor required varies depending on the type of farm. Large-scale fish farms require higher labor compared to small-scale farms. Labor is needed when feeding fish. It is also required when pumps have to be switched on and off unless the pumping process has been fully automated (Edwards, 2015). When measuring oxygen levels and the pH levels of water, human labor is needed. It is also needed when the water has to be treated and purified. The harvesting, transportation and sale of fish products also require human labor. Depending on the number of laborers required in the fish farm, the farmer will have to incur costs depending on the agreement between the farmer and the worker. However, if the worker offers highly skilled work then their pay will have to be higher compared to costs incurred to those who offer lower skilled labor.

6.4 TREATMENT OF DISEASES

One of the areas where farmers have to incur relatively high costs is when there are disease outbreaks and when fish have contracted diseases. Fish diseases pose a big threat to farmers as it increases chances of farmers incurring losses due to death. The most dangerous situation is when the disease in question is very contagious and can easily spread throughout the farm and wipe out the fish population. In cases where the disease affects certain species of fish then the situation can be managed as the farmer can be able to identify and seclude affected fish.

Figure 6.13: Outbreak of diseases could lead to an increase in production costs.

Source: https://www.fitzfishponds.com/blog/fish-disease-spotlight-ichthyobodo/

When it comes to handling fish diseases, relatively high costs may be incurred as various costly processes have to be followed. For instance, if there is an outbreak of disease then the farmer will have to pay for the services of a veterinary doctor specialized in handling fish. The doctor will have to evaluate whether the fish has been infected with the disease. The same procedure is followed when the farmer suspects that certain fish species are ill. The doctor will then have to obtain samples of water as well as that from the fish and take them to the laboratory for testing. The farmer will have to incur the cost of testing the samples in a laboratory. Once the results have been obtained from the laboratory, the farmer gets the diagnosis. If the results show that the fish do not have any disease then they will be saved from incurring further costs. However, if the fish have been diagnosed with the disease then necessary treatment protocols have to be taken up.

If the disease is very severe, the farmer will have to start over as he will have to dispose of fish species that have been affected. However, if the disease can be treated then the farmer will have to deal with the cost of purchasing the necessary medication. The cost of the medication will vary depending on the type of disease the fish was affected with. There are certain diseases that require very expensive medication while others can require the administration of locally available medicine. Farmers will also have to bear the cost of vaccinating fish from various diseases.

Other costs that could be incurred by farmers include marketing. This is when the farmer produces and sells the fish products on their own. In this case, they will have to use various marketing strategies some of which will require the farmer to incur additional costs. In some countries, the farmer will have to incur the cost of acquiring a license. The license will have to be renewed over a given frequency at a certain fee. There may be other additional costs depending on the type of activities being done on the farm.

Chapter 7

Challenges and Advantages of Extensive Aquaculture

CONTENTS

7.1 INTRODUCTION

Figure 7.1: Aquaculture.

Source:https://www.bluemarinefoundation.com/wp-content/uploads/2019/10/Aquaculture-02.jpg

Aquaculture is defined as the cultivation of aquatic organisms such as fish as well as other aquatic plants. This type of agriculture is rampant in freshwaters and salty waters where the fish are harvested in a controlled amount to avoid overfishing that is bad for the general environment. Another type of farming that can be compared to aqua agriculture is commercial fishing where wild fish are harvested. On the other hand, modern agriculture is the practice of agriculture in seawater habitats as opposed to fresh aquaculture. Aquaculture is mainly conducted in artificial facilities which are built on land and this is known as onshore aquaculture. The survival of fish solely depends on human control where people are assigned tasks to ensure that the fish are well taken care of. For the inshore aquaculture, the fish are left in a natural environment and they depend on natural conditions such as ocean currents and nutrient cycles.

There are certain interventions in the fishing process which are meant to enhance the production of fish and some of these interventions include regular stocking, feeding the fish, and protecting the fish from predators. They are various forms of aquaculture and they include the obvious one which is farming of fish, oyster, mariculture and algaculture which is basically the farming of seaweed and cultivation of ornamental fish. There are two methods of farming that integrate fish farming and aquatic plant farming. These methods are called Aquaponics and integrated multi-trophic aquaculture (Díaz et al., 2019). According to the Food and Agriculture Organization, aquaculture is considered to be one of the industries that are directly affected by climate change and experience most of its impact. However, it is important to note that some forms of aquaculture also have a negative impact on the environment because they pollute nutrients and diseases are transferred to wild populations.

7.2 CHALLENGES OF EXTENSIVE AQUACULTURE

Figure 7.2: Challenges facing extensive aquaculture.

Source: https://image.slidesharecdn.com/iafl302fl0ds-130328052836-phpapp02/95/challenges-facing-the-aquaculture-feed-industry-2-638.jpg?cb=1364448580

There are a number of benefits that aquaculture brings to the environment however, there are certain challenges faced by aqua agriculture and they make the farming of aquatic animals and aquatic plants hard for the farmers who practice this type of farming. It is important for people looking to practice aqua agriculture to understand the challenges that face extensive aquaculture or sores that are brought about by extensive aquaculture.

7.2.1 Environment

One of the challenges facing aquaculture is the environment. As stated this type of farming is usually done in an environment where humans are able to control the different variance that affects the development of the fish. Having this in mind, land fish farming which is mainly conducted in tanks requires a lot of supervision especially because the water found in those tanks has to be changed frequently in order to avoid abnormal development of the fish. When cleaning the tanks the environment can be polluted because the water containing discharge from the fish and chemicals are usually released to the environment that therefore affects the surrounding and those people living around the areas where farming of fish is practiced. Once these discharges are released into the environment there is a rampant growth of algae, which eventually removes the oxygen that has already dissolved in water that can also be termed as eutrophication.

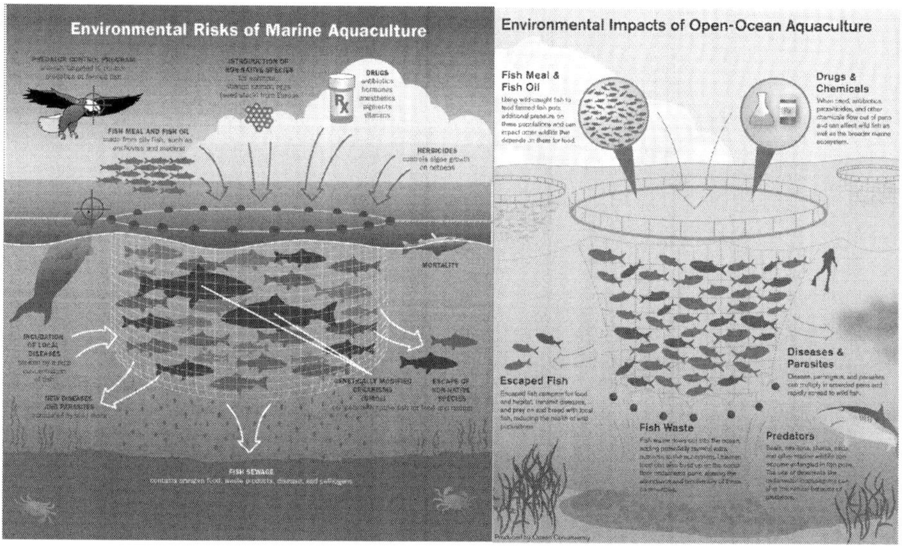

Figure 7.3: Environment as a challenge of extensive fish farming.

Source: https://hydroheart.org/wp-content/uploads/2020/09/3092017_orig.jpg

It is also important to note that marine fishing, crustaceans and even bivalves are known to produce waste products in the form of fecal matter as well as unused feed. These waste products are known to contain high levels of nitrogen which cause oxygen depletion especially in coastal environments which results in loss of marine productivity in most coastal areas. As stated another issue is the use of antibiotics, antifoulants, and certain pesticides which are introduced to the marine environment and cause irreversible damage to both marine life and those areas that surround the marine environment (Diana et al., 2013). When oxygen is removed from the water, fish are unable to sustain because they need oxygen so that they can live. There are some chemicals used in aquaculture that are harmful to the environment and the fish. Some of the antibiotics and water treatment agents that are used in this type of farming are usually released into the waterways that leads to pollution. In this case, to avoid cases of pollution the aquaculture systems have to be closed or the waste that is released to the environment needs to be treated before being discharged.

7.2.2 Disease Spread from Aquaculture Farms

Sl. no.	Bacterial disease	Causative agent
1	Vibriosis	1. *Vibrio* spp. including *V. anguillarum, V. harveyi clade, V. parahaemolyticus, Aliivibrio salmonicida (V. salmonicida), V. vulnificus*
		2. *Photobacterium damselae*
2	Aeromonasis	*Aeromonas* spp. including *Aeromonas caviae, A. hydrophila, A. sobria, A. veronii, A. jandaei, A. salmonicida*
3	Edwardsiellosis	*Edwardsiella anguillarum, E. ictaluri, E. piscicida, E. tarda, Yersinia ruckeri*
4	Pseudomonasis	*Pseudomonas anguilliseptica, P. fluorescens*
5	Flavobacteriosis	*Flavobacterium branchiophilum, F. columnare, F. psychrophilum, Tenacibaculum maritinum*
6	Mycobacteriosis	Caused by *Mycobacterium* spp. and other filamentous actinobacteria:
		1. *Mycobacterium fortuitum, M. marinum*
		2. *Nocardia asteroides, N. crassostreae (ostreae), N. seriolae*
7	Streptococcosis	*Streptococcus agalactiae, S. iniae, Lactococcus garvieae, Aerococcus viridans*
8	Other bacterial diseases	1. *Renibacterium salmoninarum*
		2. Anaerobic bacteria such as *Clostridium botulinum, Enterobacterium catenabacterium*
		3. Intracellular infections by *Piscirickettsia salmonis, Hepatobacter penaei, Francisella noatunensis, Chlamydia* spp.

Figure 7.4: Infectious diseases in aquaculture (here few bacterial diseases of aquaculture animals are presented).

Source: Jayaprakashvel, Mani & Subramani, Ramesh. (2019). Implications of Quorum Sensing and Quorum Quenching in Aquaculture Health Management. 10.1007/978-981-32-9409-7_18.

The operations associated with this type of farming have been noted to cause the spread of parasites and diseases into the wild. even livestock sharing can cause the spread of these parasites in the scissors into the wild and therefore finding fish should be conducted in areas where there is close supervision to avoid pollution that is dangerous to the environment. There is an increase in the chances of getting parasites such as sea lice in farmed fish compared to those fish that are bred in natural environments. The farmed fish are exposed to diseases through the use of unprocessed fish as a food source. Most farmers will prefer to use unprocessed food fish because they are cheaper as opposed to the recommended safe processed fish pellets. When the diseases and parasites spread from the fish into the wild they cause irreversible damage to the general environment. Some of the diseases spread are usually very severe and can cause loss to the people practicing different types of farming.

Farming in the world environment can be a vector for disease spread. As stated salmon fish tend to escape from their habitat, which increases the transfer of diseases. The transfer of these diseases is usually very rampant and is considered to be a great phenomenon that affects the entire environment. One of the diseases spread by salmon is called infectious salmon anemia. This disease was first reported in Chile during the 1990s and continued to spread across the world. It has been noted that this disease has continued to spread faster than any other disease because of various factors. Some of the factors include poor biosecurity which means that salmon is not well protected and taken care of which results in the spread of the disease. Another factor is a poor global transfer of salmon that has accelerated the spread of the infectious disease is salmon anemia from country to country and even it has been reported that this disease is spread from one continent to the other (Diana, 2009).

Sea lice are another side effect of aquaculture that is noted in agriculture and is from the genus Lepeophtherius and Caligus. The sea lice are usually dangerous because they attach themselves to the body of the intended victim and then they go ahead to draw nutrients from the body, therefore, compromising the immunity of the victim. Those who criticize aquaculture have noted that large operations established near the migration routes of wild salmon can cause sea lice to aggregate and jump from farmed animals to the wild ones.

7.2.3 Escapees

Foreign species are usually introduced into new areas because of aquaculture. Under the right conditions, the introduction of different species can create an unhealthy spread of invasive species. Animals that have escaped their natural habitat such as farmed fish and animals which are put in pens can cause damage to the environment and are capable of threatening native fish populations. Escape of farm fish can cause a lot of damage because they compete for food and habitat which results in the displacement of indigenous ones and interferes with the life of wild species. It is vital to note that these fish can be carriers of diseases and parasites that will eventually kill other native species and interfere with their general environment. The natural gene pool is threatened because these escapes result in breeding with other livestock to cause the dilution of the gene pool that is needed for survival. This, therefore, creates a threat to the long-term survival and evolution of wild species. The addressing of escapes or found fish has failed until recently. The escape of fish is usually rampant in salmon aquaculture. Most of the salmon fish usually escape from net cages that are usually suspended in the open Ocean. Tilapia, which are grown in fresh ponds, have a great chance to take on a watershed.

7.2.4 Habitat destruction

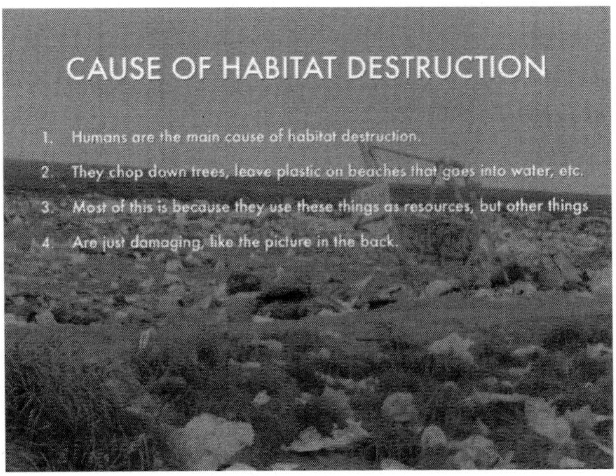

Figure 7.5: Habitat destruction.

Source: https://img.haikudeck.com/mg/E79EE0B6-D74D-49E3-9BBC-05502E547543.jpg

Out of all agricultural practices, the farming of shrimp has faced many criticisms. Shrimp farming is mainly conducted in tropical and subtropical ponds which are usually contained in mangrove forests in coastal regions. It should be noted that the accumulation of pollutants is rampant in ponds which led to the abandonment of those ponds and farmers seek better areas to set up their farming activities. There are other factors that led to the abandonment of these ponds such as diseases caused by bacteria and viral "white spot". This process resulted in the destruction of many mangrove forest ecosystems, which were Whitehall introduction of wild Fish and the protection of the coast from storm surges. Once the mangrove forests were done away with it became easier for pollution to happen and the storm surges led to extensive damage especially to those people who lived in the areas (Diana, 2012). Over the years the destruction of mangrove forests has become significant but in some countries mangrove cutting has become illegal. However, it is important to note that despite these changes new challenges continue to arise and are facing marine farming. In 2009, Early Mortality syndrome was abbreviated as EMS arose in China and continued to spread throughout Southeast Asia. According to statistics the losses that were incurred as a result of the Early Mortality syndrome one more than $1 billion.

7.2.5 Fish Diets

Most of the fish which are farm-raised are usually carnivorous. However, there are certain types of fish that are considered to be omnivorous and they are shellfish and shrimp. The carnivorous type of fish includes trout, salmon and Kampachi. During the 1970s and 1980s, most farmers opted to fish as farmed fish with diets that were rich in Fish oiled and Saturn fish meals which were acquired from wild fish. Some of these wild fish included sardines and anchovies. The diet given to the fish was considered to be nutritional. However, this was very expensive for the farmers because five kilograms of good wild fish was given to only one kilogram of farmed fish. Over the years as aquaculture grows the demand for fish oil and a meal gotten from wild Fish has continued to increase but the problem is that the supply is not efficient to ensure that every fish is well fed. The USDA is looking to find protein and vegetable proteins that are suitable for farmed fish in order to address the issue of lack of enough food for the farmed fish. Fish oil has continued to be a bigger problem than fish meals over the recent years. Fish oil is particularly important because it contains omega-3 fatty acids that are needed by the fish to ensure that they develop into strong organisms that can withstand natural conditions.

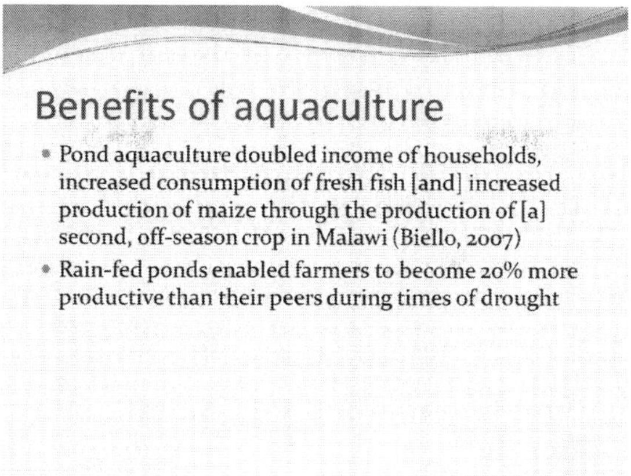

Figure 7.6: Benefits of aquaculture.

Source: https://i1.wp.com/eschool.iaspaper.net/wp-content/uploads/2018/04/Benefits-of-aquaculture-300x225.jpg?resize=300%2C225

7.3.1 Growth of the Economy

According to the research conducted, aquaculture is considered to be a multibillion dollar industry. The money that is generated from this type of funding is usually enough to facilitate the funding for various activities in different states, cities and also the national government of different countries. It should be noted that there are a lot of people who continue to consume fish throughout the nation. There are various markets where fish is most traded. These markets are found in places such as Alaska, the New England area, the gulf coast region, and the Northwest Pacific region. These areas are considered to be the top seafood consumers in the United States. There are other businesses that depend on aquafarms such as restaurants that serve seafood grocery stores and other fish markets (Cressey, 2009). Most aquafarms usually use biofuels and other materials for fish food. This type of farming has led to the economies of various states becoming strong and productive because of the money that is sourced from these types of farming. Many countries have continued to venture into this type of farming as a way of sustaining the livelihood of those who live in the countries and encourage development which is a paramount priority for most countries across the world. The colleges and universities that are located near aquafarms are usually important because they act as educational, and research facilities for students who are interested in venturing into aquafarming.

7.3.2 Acts as a source of food for most families

This type of farming is beneficial to most families because it acts as a source of nutrition for millions of people across the world. Aquafarming helps to supplement the food supplies of various countries such as the United States. Aquafarming that is done in farms acts as a source of meat for most of the American diet. However, it is vital to note that some regions found in the United States of America do not have a standard means of ensuring that meat products are made available for the American diet. This has continued to be an issue over the years because not everyone in the US is a vegetarian and accessing healthy meat products especially white meat is usually very difficult and leads to a lack of certain nutrients that are needed by the body for normal development. In this case, aquaculture is important because it acts as a supplement for people who lack meat products. The different fish species are used to feed the animals and people which is considered to be a win-win situation for most people because it saves on money. According to the statistics collected, more than 13 million people who reside along the American coastlines depend on aquafarming as a source of food supply.

7.3.3 Aquaculture helps protect various coastlines and aquatic environments

The coastlines that are found in the United States of America are usually affected by the issue of pollution. Water pollution is a new issue that has continued to grow over recent years and has affected various aspects of people's life. People tend to dispose of waste products that are harmful to the organisms living in water and have resulted in the decrease of both aquatic plants and animals. There are cases where some of the nation's water is usually too polluted to sustain any form of life. Aquaculture is important because it helps to reduce waste within marine environments (Chavanne et al., 2016). However, in order for fish farming to be successful, the areas where the farming process is being conducted have to be free from pollution that is associated with reckless disposal of garbage. Farmers who practice aquaculture help to protect the environment by ensuring that the coastlines are free from any type of pollution that might interfere with the feeding of fish and other aquatic plants such as the lotus. They ensure that the marine environments are well-maintained as a way of offering concrete support for their fish species and other forms of wildlife in the surrounding area. It is vital to note that aqua farming must be well-maintained to create high-quality marine food for consumption.

7.3.4 Aquafarming helps to protect species and habitats

Due to the pollution that goes on in the marine environment some of the species are in danger of being eliminated. In most situations, the marine habitats are also in danger of being destroyed because of various issues such as overfishing and overpopulation of one species of animals which act as predators and prey on other marine life to an extent of extinction. Another issue is that dying fish species are disappearing from the environment and this causes the likelihood of the majority of the marine environment to be eliminated. On the other hand, the species habitats are also destroyed as a result of pollution and other specific factors. Some of these factors include the clearing of land as a way of encouraging development or increasing other farming activities and cases of people who destroy their habitats because of carelessness or intentional actions. Fish farming helps companies and other farmers to ensure that their environment is well taken care of as a way of minimizing pollution and helping to protect species and their habitats. It is also considered to be a way of regulating wildlife species by stopping the population of some species and eliminating other species which cause more harm than good. This specific process ensures that the habitats are healthy and strong to help sustain aquafarming. The selective breeding raising is also used in Aqua funding to help with the introduction of species that were extinct to an environment and help improve the numbers of existing species, which are in danger of extinction.

7.3.5 Aquaculture is important for scientific research development related to marine life

There are people who utilize aqua funds as a way of studying different fish species. Most of these people include Marine biologists, oceanographers and other various marine scientists. These people conduct research as a way of ensuring that various fish species are being raised in a healthy environment, which encourages proper and healthy development of the fish. They are also responsible for examining how certain species of fish interact and grow within a selected environment (Cole et al., 2009). Some of the other research usually involves cross-breeding various types of fish species and keeping a record of how the different breeds of fish usually interact with each other. These people are also tasked with evaluating the environmental impact of various fish species. The scientists are responsible for monitoring the soil where the aqua farming is done, ensuring that what is used for this type of funding is either fresh or salty and in evaluating the environmental conditions which are used to determine how this type of farming actually impacts the environment in general. Marine scientists usually use their research as a way of helping them to figure out how to come up with fisheries that are highly sustainable and profitable.

Figure 7.7: Global benefits of aquaculture.

Source: https://i.ytimg.com/vi/2NHVHHSYxUs/maxresdefault.jpg

Chapter 8

Development of Extensive Aquaculture

CONTENTS

8.1 INTRODUCTION

Aquaculture rehearses have a long history, however, aquaculture as a food production area on a worldwide scale is generally juvenile. The area is developing consistently and as indicated by FAO, the aquaculture area addressed 53% of the absolute fish production from fisheries and aquaculture in 2015.

Aquaculture production has continually expanded its offer in the absolute fish production, while the yearly catch from fisheries has remained practically unaltered since the 90s.

Fish utilization has been developing and the additional interest has been met by the aquaculture area.

The heightening of aquaculture to item more sea-going living beings, without utilizing numerous assets nor hurting the climate, is one of the principal challenges that the business faces. This article plans to create a structure that characterizes heightening according to a production viewpoint.

The beginning stage of the current examination is a formerly created guide of aquaculture production systems dependent on an efficient writing audit. The motivation behind the guide is to give an outline of the creative capacities expected to cultivate sea-going animals in aquaculture systems. The guide likewise gives an outline of the techniques used to complete them and the primary boundaries used to gauge each capacity (Costa-Pierce et al., 2021).

This paper then, at that point expands that work into meanings of aquaculture power dependent on what production capacities are completed. The arrangement is to make a rational structure that expressly characterizes extensive, semi-intense and intensive aquaculture for every fish organic entity.

The definition depends on production capacities, yet is still following the current utilization of the terms. The current utilization of the terms is in no way, shape or form intelligible, infrequently characterized unequivocally, not utilized in a similar way, identified with various parts of sea-going species cultivating and vary between what types of living being will be being refined.

This is evident in the majority of the writing utilized in this article with genuine models some scholarly work.

Figure 8.1: The USDA National Aquaculture Research Center in Stuttgart, Arkansas.

Source: https://edis.ifas.ufl.edu/publication/FA217

8.2 EXTENSIVE AQUACULTURE SYSTEMS

As generally distinguished in the writing, the most extensive systems are those where there is pretty much nothing or even no human obstruction. As a result, those systems by and large produce not exactly those of greater force. A typical sort of extensive framework is the place where a confined zone made by a net, enclosure or some kind of a fence is embedded in a bigger water body where animals can be refined inside. Another sort can be a lake farm where no extra taking care is utilized and the biological system inside the lake gives feed to the refined animals. Extensive systems don't appear to have acquired a lot of consideration deciding on the material dissected for this audit. That ought not to come as a shock considering the way that those systems apply a base measure of capacities. One scholar expressed that extensive aquaculture systems looked like the regular habitat of the occupants without applying supplemental food. Another characterized extensive aquaculture as softly supplied systems where water throughput isn't supported and feed or manure inputs are not applied.

Experts gave a comparable definition and alluded to extensive systems as those relying upon normal food sources inside the way of life unit where feed extra feed inputs are not added purposefully. By the by, extensive systems have been associated with capacities like taking care of or treating in the writing (Colsoul et al., 2021). This demonstrates an irregularity in the definitions of the force of aquaculture systems. As per how extensive systems are by and large portrayed in the writing, we accept that extensive systems apply the base measure of capacities needed for an aquaculture framework to be practical. Obligatory capacities are set apart as 'required' while discretionary capacities are shown as 'discretionary'. In extensive systems, the yield capacity of reaping is constantly applied and subsequently checked 'required'. The elements of providing water, loading, taking care of, treating, and forestalling gets away should be applied now and again. The yield capacity of 'forestalling departures' is additionally once in a while utilized. Other information and yield works that are not shown in the guide were never connected to extensive systems in the articles inspected. The equivalent goes for treatment capacities.

Figure 8.2: Recirculating aquaculture at the UF/IFAS Indian River Research and Education Center.

Source: https://edis.ifas.ufl.edu/publication/FA217

8.3 LAKES, WATERWAYS AND SUPPLIES (EXTENSIVE AQUACULTURE)

Business freshwater aquaculture in the Mediterranean nations began toward the finish of the last century. Coldwater salmonids, chiefly *Oncorhynchus mykiss* and Salmo trutta, are the main gathering among the aquaculture production of France, Italy, Spain and Turkey. Salmonids are as yet thought to be a centre high value item, popularized, new or prepared, bound for their homegrown business sectors, yet additionally traded. The high volume of this production is likewise its significant limitation in light of the reformist immersion of the interest. A lethargic expansion underway is, all things considered, still anticipated.

Warmwater freshwater species (initially dependent on normal carp, to which presented Chinese carps and tilapia were subsequently added) address the significant aquaculture production in Croatia, Bulgaria, Romania, Israel and Egypt and are completely consumed- locally. In the three European nations of this gathering the production declined forcefully because of the political progress time frame, and as of late showed a positive pattern once more (high potential was shown for this kind of aquaculture was accounted for in these three nations). Despite what is generally expected, in Israel and Egypt production has consistently expanded and today contributes fundamentally to public food security and limits the import of fish. In these two nations, the fundamental imperative is the restricted accessibility of freshwater (clashes with horticulture and prerequisites for homegrown water utilization) so both are attempting to expand the stocking density in this structure of fish raising.

Figure 8.3: Split ponds for ictalurid catfish in Mississippi, USA. The middle pond is highlighted to show the partitioning of the original 3.2-ha pond into a 0.6-ha fish-holding basin and a 2.6-ha algal waste-treatment lagoon. Arrows show the direction of daytime pumped-water circulation through culverts.

Source: https://onlinelibrary.wiley.com/doi/10.1111/jwas.12714

Marine aquaculture started all the more as of late, during the 1960s, and has shown quick development. The high item value of the species raised (except for mussels) coordinates this item towards trade showcases yet, simultaneously, marine aquaculture is likewise seen by numerous nations as a way to restrict fish imports. The area has been for quite a while overwhelmed by two species, ocean bass (*Dicentrarchus labrax*) and ocean bream (*Sparus aurata*) at the same time, during late years different Sparids (*Diplodus sargus*, puntazzo) and, all the more as of late still, the imported *Sciaenops ocellatus* (Israel) are adding to the extensive ring of production. Mollusc culture is all the more usually a corporate movement, bringing direct certain social advantages, while fish culture is for the most part overwhelmed by private undertakings. Shrimp culture is as yet a minor area and green growth/macrophyte development are essentially missing in the locale. Marine aquaculture is dealing with normal issues in every one of the nations, like a reformist immersion of accessible destinations (both for extensive and intensive aquaculture), high rivalry in coast regions use (particularly with the travel industry improvement), and market limitations (especially because of the new EU import guidelines). A somewhat sluggish yet constant development of my aquaculture area is for the most part expected, while seaward fish culture is the movement that is drawing in likely financial backers (Bjelland et al., 2015).

There are two significant exercises in this area: i) re-loading for business fisheries and ii) cultivating for sport fisheries. The two exercises are typically upheld by open specialists managing fisheries and aquaculture, with ranger service, or with agribusiness (water system). The area is just pertinent in a couple of nations (Bulgaria, Egypt, Romania, Tunisia; a few great many tons each year), and exercises happen in a wide scope of water bodies, which contrast in size and use (for the most part supplies utilized for power production or water system), which support a colossal number fishermen either having a place with cooperatives or working exclusively. Clients now and again have elite fishery rights and just in a couple of cases do they need to pay for the asset use (by renting the water surface or paying for the fingerlings presented). Fishing people groups are once in a while straightforwardly associated with fingerling production as well as the board (Egypt, Morocco, Romania).

The most widely recognized species supplied are rainbow trout (*Oncorhynchus mykiss*), normal carp (*Cyprinus carpio*) and the presented Chinese carps (*Ctenopharyngodon idella*, *Aristichthys nobilis*, *Hyphophtalmichthys molitrix*). Different types of Tilapia, pike (*Esox lucius*) and roost pike (*Stizostedion lucioperca*) are additionally regularly utilized. Egypt is a unique case in that it effectively rehearses culture-based fisheries in inland saltwater (Lake Quaroun) by utilizing marine fish and shrimp species. Wild-caughtfingerlings of grey mullets (*Mugil cephalus* and *Liza ramada*) are utilized to restock supplies in Tunisia, which has added to an increment in freshwater fish production from 700 t in 1996 to 1,010 t in 1997. Be that as it may, production is profoundly subject to yearly variances in the inventory of wild-caughtfingerlings. Re-loading lakes and streams on the side of calculating are normally done in most Mediterranean nations. Fishermen pay for fishing licenses.

Extensive inland aquaculture, including society-based fisheries, shows differentiating patterns. A sharp diminishing has described the area during the previous decade in Bulgaria (legitimate and monetary limitations, because of the political change time frame) and in Cyprus (given the new dry season), while the restricted accessibility of fresh water doesn't permit its development in Israel. Despite what is generally expected, it is quickly growing in Egypt and is accounted for as having high potential in Morocco. In the excess nations, the social effect of this action is somewhat restricted and production from extensive aquaculture stays stable at a low degree of production or is restricted.

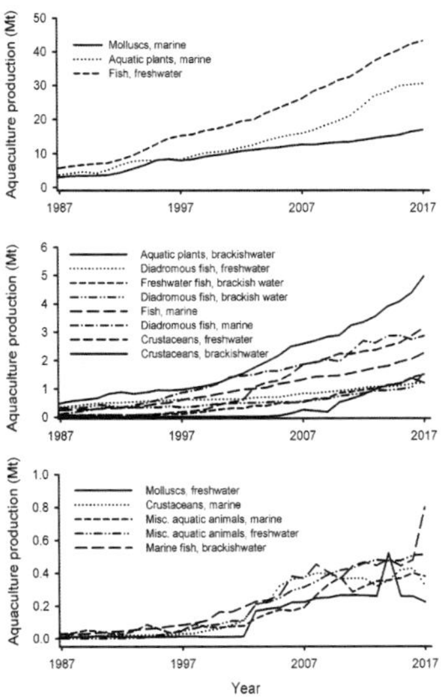

Figure 8.4: Global aquaculture production (m.t.) separated into three magnitudes of volume and by species groups and aquatic environment.

Source: https://onlinelibrary.wiley.com/doi/10.1111/jwas.12714

8.4 CONDITIONS NEEDED FOR THE START OF NEW EXTENSIVE AQUACULTURE SYSTEMS

8.4.1 Getting Good Quality Juveniles

Albeit the little size of the eggs of the incredible greater part of fish species confuses artificial generation, incubation centre procedures have worked on impressively in the course of recent years. In aquacultural exploration, the authority of this capacity has regularly been viewed as a need, regardless of whether artificial proliferation is not generally fundamental for fire up extensive development (cf. shellfish culture or the transplantation of juvenile place). Today the scope of species whose generation is well or genuinely very much dominated is moderately enormous (Appendix 2) and given present information and ability this rundown could be reasonably quickly reached out later on. The production of juveniles is in this manner not a significant resistance to the advancement of extensive aquaculture.

Then again, the interest in fake propagation (consistency of the item, decrease of normal perils, less reliance on seasons and hereditary determination) develops with the improvement of extensive systems. Regardless of whether the potential applications are less enhanced than in serious development, the hereditary choice may add to the battle against epizootic illnesses, further develop development execution, or serve to stamp the people delivered (to assist with assessing endurance rates or hereditary blending in with wild stocks). Be that as it may, choice additionally makes chances for the uprightness of the hereditary legacy of regular populations. As of now, the absence of fundamental information about the hereditary qualities of certain scientific classifications, like mollusks, is a requirement for determination (Bjørndal, 2002).

Figure 8.5: A "Taiwan-style" paddlewheel aerator on a fish farm.

Source: https://onlinelibrary.wiley.com/doi/10.1111/jwas.12714

8.4.2 Forcing Enrollment

For the majority of the species which have been the subject of delayed experimentation there is plentiful writing portraying:

- the characteristics that the delivered people should have: beginning of the stock (regular or incubation centre juveniles, hereditary legacy as a component of the environment to be colonized), size and age, transformation to the life in the normal natural surroundings (on account of salmon, for instance, earlier obtaining of the reflexes and practices of taking care of in the wild and protection against predators);
- the conditions for the arrival of people: environment, date concerning the smolt stage for salmon, occasional trophic advancement of the environment, event of predators, and so forth, with, as a potential culmination, activities to set up the site (end of predators like starfish in scallop development in Japan, for instance).

However, in any event, for solitary animal types, the ends are not in every case exceptionally concordant. This is clarified by the zootechnical approach frequently liked in R&D programs. Without fitting information on the biology of juveniles and the propagation systems of populations in their current circumstance, it is hard to characterize the characteristics that the alevins need to show and the conditions important for their delivery (Bottema et al., 2021).

The requirements of fish life - and specifically migration - force complex segment methodologies on marine populations. These techniques display various normal attributes in any case:

- extremely high fruitfulness, to make up for the colossal mortality endured by the eggs and hatchlings whose method for movement and food saves are restricted;
- the reliance of every population on a progression of hydrodynamic designs and the choice of physiological cycles and practices adjusted to the conditions getting in the constructions involved throughout the progressive phases of the existence cycle (eggs, hatchlings, juveniles and grown-ups);
- homing conduct: to finish their life cycle and guarantee the endurance of the population, and not just of people, an adequate number of reproducers need to return occasionally to foreordained

spots in their living srate. The re-visitation of fry lakes that are steady in reality along these lines gives off an impression of being a conduct reaction to the twofold basic of sexual multiplication and life in the oceanic environment.

Figure 8.6: The full-size (7.5 kW) prototype paddlewheel aerator based on research conducted at Auburn University.

Source: https://onlinelibrary.wiley.com/doi/10.1111/jwas.12714

These imperatives imply that marine species are comprised of populations which are recognized by the progression of examples of dispersion and relocation explicit to the various phases of the existence cycle, their segment attributes (normal plenitude, development, regular mortality), and their hereditary legacy. During the underlying stages, the numbers in every population fall significantly under the impact of natural cycles autonomous of the plenitude of the population (impact of temperature on sexual development, gametogenesis and practicality of the eggs as on account of the scallop; floating into unfriendly hydrodynamic designs where the juvenile stages may either die or endure yet lose any shot at taking an interest in the generation of the population, as on account of a few types of fish), and trophic cycles relying upon the density of the population (predation, starvation).

These segment techniques have vital ramifications for the plan of top-notch research programs:

- as the guideline measures are gathered in gametogenesis and the underlying phases of the existence cycle and are, essentially somewhat, free of density, it is truly plausible that for decent numerous animal varieties the biotic limit of the territory involved by the grown-ups is certifiably not a restricting element for the degrees of enlistment generally noticed for normal populations; consequently significant overpopulation ought to be viable with the trophic limit of the environment. This theory is affirmed by experience (shellfish culture, salmon farming), just as by the extraordinary plentifulness of normal enrollment in specific species;

- from the rigorously biological point of view, the overall idea of homing conduct and the way that a greater part of marine species imitate in seaside regions permits us to presume that ocean farming ought to be feasible for an extraordinary assortment of animal types;
- the specialized disappointment of numerous test programs is most likely clarified by deficient considering of the particularities of the multiplication procedures of marine populations; gigantic mortality happens when the people delivered don't discover defense (lobster, abalone), or the food they need, or succumb to normal predators (scallop in France, lobster and cod in Norway, and so on);
- the decision of the stock is frequently basic for the achievement of restocking and acclimation programs: when in doubt, the achievement rate for salmon farming tasks falls when the restocking is made with stocks coming from another sea. The stocks must be adjusted to the length of the waterway or stream to be capable swim up to it. Certain poor rates of return are clarified by a poor procurement by developed animals of homing, a part of which is most likely genetic. The proliferation methodology of the scallop, for which the accomplishment of enlistment relies upon synchronization among gametogenesis and the late spring warming cycle in the natural surroundings involved by every population, could well motivation restocking with spat from populations living in totally different territories to be a disappointment; (Bentzon-Tilia et al., 2016).
- Then again, as the two juveniles and grown-ups of assorted species may make due in hydrodynamic constructions not the same as those occupied by the populations to which they have a place, and are more lenient than eggs and hatchlings, the biological requirements ought to by and large be less extreme for buffering and overpopulation tasks not including the support of the delivered people in proliferation. Shellfish development in areas extremely far off from their srate of generation is reliable with this theory.

Figure 8.7: An intensively aerated biofloc shrimp pond in Guatemala.

Source: https://onlinelibrary.wiley.com/doi/10.1111/jwas.12714

As the biological imperatives change with the targets, the achievement of the R&D programs hence relies upon the express meaning of the goals, the hidden theories and the test conventions. These conditions are hard to meet in projects where innovative work are not recognized. With signs of progress in information about the generation techniques of marine populations and the interest displayed by marine biologists in the full-scale natural investigations comprised by projects to foster extensive systems, just as in the opportunities for experimentation on the environment and physiology of propagation offered by incubation facilities, these angles are presently better considered in extensive aquaculture programs (scallop in France, Norwegian PUSH program on cod, and so on).

8.4.3 Availability of Space

Rivalry for locales with effectively settled employments of the marine strip might be a hindrance to the implantation of new fish farms. Shellfish culture at first created on the intertidal zone, in straights and bowls whose shores were not especially pursued at that point. In nations where shellfish culture has been embedded for quite a while (Spain, France, Netherlands, and so on), a decent extent of the great locales has now been colonized. A restricted augmentation of development towards the vast ocean stays conceivable because of specialized advancements (clams development in profound water, mussel culture on nets). Where it is grounded, shellfish culture profits by the benefit of its anteriority in the allotment of new concessions. This isn't the situation with new implantations (mussel culture) which regularly face the resistance of nearby occupants. Ecological assurance contentions might be advanced with more prominent power against new types of development than against traditional utilizations (fishing, agribusiness). These practices uncover the shortcoming of current strategies for apportioning the oceanic space between contending employments. The way that few sorts of extensive development (migratory species and sessile species) make moderately little interest in the oceanic space could be a benefit in areas where the seaside strip is exceptionally pursued.

8.4.4 Biotic limit of the environments and water quality

While, for the reasons clarified above, the biotic limit of the environment is for the most part not a hindrance for the beginning up of new extensive systems, this perception is certainly not an outright principle. The usefulness of the biological system isn't adequate wherever to allow development rates viable with beneficial shellfish cultivating. Urban and agrarian users may, inside specific cutoff points, cure this destitution, similarly as certain tidy up projects may lessen the efficiency of certain shellfish cultivating bowls (Amenyogbe et al., 2018). More basic for the advancement of extensive aquaculture, and all the more especially of shellfish culture, is the nature of the water. Taking care of through filtration and bioaccumulation in the natural pecking order make shellfish cultivating stocks and items entirely powerless against any damage like coast waters under the impact of urban, agrarian and modern pollution.

8.4.5 Technical Productivity and Benefit

To assess the specialized productivity of another framework, it is important to have the option to distinguish the people when they are recovered. A few checking strategies are utilized for this reason: outer and inward imprints, genetic imprinting (Norwegian PUSH program for cod), or morphological imprinting (in Japan, for instance, incubation items may have trademark qualities: the unique type of the nostrils of ocean bream, dark pigmentation of the visually impaired face of wallow, and so on) The chance of changes in the principles of admittance to the stock and the forcing of a commitment to the financing of alevin raising may make fishermen under-announce the recovers. With great insights, the assessment of the effect of raising projects on the biomass of a population and the fishing yield doesn't present any specialized troubles. Then again, the isolation of the various reasons for mortality and losses (regular mortality, overabundance mortality of the artificial stock, migratory rate, catches by various armadas) may demonstrate troublesomely.

Figure 8.8: The benefits and problems associated with aquaculture.

Source: https://ecosystemsunited.com/2016/11/27/the-benefits-and-problems-associated-with-aquaculture/

The presence of a steady deficiency of ocean results in high business value in the rich nations and the way that the advancement of extensive techniques isn't by and large any more hard for respectable species than for others are factors good for the improvement of extensive aquaculture. Then again, the poor transport, regular in extensive systems, among social and private expenses and advantages is a significant debilitation. It is the essential driver of the poor effectiveness of numerous extensive aquaculture programs, and of the regularly insuperable challenges experienced in attempting to move from public to private financing. Exhibition of the specialized and monetary suitability of another framework infrequently does the trick to guarantee its take-off.

All together for a program to be beneficial, it does the trick on a basic level at the cost of the fish recovered to surpass the expense of production and recovery of the people delivered. The deficiencies of the rejection systems imply that profits by outsiders must be considered losses (notwithstanding those because of normal mortality and the wandering of people), while from the specialized outlook they compare to gains. A not inconsequential extent of both the additions (benefits for the preservation of piscicultural fauna) and the expenses (utilization of the normal biotic limit, the hazard of genetic mixing, adjustment of the scene, and so forth) are social (Allsopp et al., 2013). The financing will be even harder to move to the immediate private recipients the more noteworthy these external impacts. This is the reason a basic monetary examination at the level of the venture isn't sufficient to set up the financial and environmental value of another framework. An expense/advantage examination on the size of the program overall is fundamental. It will acquire from being supplemented by a danger examination at the level of the primary entertainers concerned (direct recipients and different clients of the environment).

8.4.6 Ownership of the Stocks

Assessment of extensive systems uncovers the presence of connections between's the monetary and social association of the activity and the cycle of strengthening like those portrayed for horticulture. All restocking programs for the preservation of the fauna are financed by the public area and advantage the general public all in all without limitation. The equivalent is for the most part valid for rearing juvenile fish to help commercial fisheries in which access remains to a great extent open and free: it is quite often financed by open assets. A fractional move towards privatization is to be found in sport fishing in which fishermen appreciate elite rights – group or individual -, as a trade-off for a commitment to the financing of the reproducing programs. Components of two systems are found in conventional fisheries: even though fishermen may contribute, the expense of the restocking programs is for the most part borne by the state, yet in traditional law, access is regularly restricted by an arrangement of aggregate regional rights. The aggregate possession system existing in Japan gives a genuine illustration of this sort of system.

Japan is one of the uncommon nations to have legitimized the advantages that marine fishermen get from their anteriority of control of the coast strip and use of seaside stocks. Altogether, fishermen' cooperatives and certain private groupings hold privileges of semi responsibility for assets. While they are not allowed to sell their privileges, neither may the state pull out these rights. This formalization of the precedent-based law isn't detached from the advancement of extensive aquaculture in this country (Allen et al., 2012). While private possession and misuse are the exemption in the extensive development of vagile species (private homesteads for use for benefit by the proprietor, or game fisheries where access is administered by singular rights), they become the standard in shellfish culture, in which the stock is promptly appropriable and successfully appropriated. The distinction doesn't end there: shellfish culture is impressively more evolved than the farming of vagile species. While the insights on this last are different, the figures accessible show that public productions are frequently a significant degree lower than those of shellfish culture. Ultimately, the private responsibility for the stock is essentially the standard in intensive fish farms.

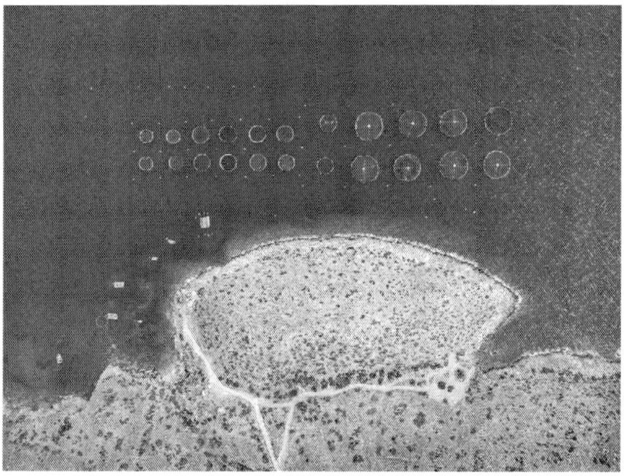

Figure 8.9: Aquaculture farm.

Source: https://www.euronews.com/green/2021/04/26/what-is-aquaculture-and-can-it-prevent-overfishing

Counter-instances of projects which, despite their specialized achievement, have not prompted business production, or whose take-off is impeded by institutional limitations, are no less clear. In France, for instance, where the extensive culture of the scallop is dominated and has been the subject of full-scale tests financed by the public area, the exchange of the expense of planting to private administrators is advancing just leisurely. The exchange associations acknowledge the way that fishermen should make a commitment, on a deliberate singular amount premise, to the financing of an aggregate restocking program, however, they are against the giving of aggregate concessions on regular beds – however, these are the most encouraging -, or to tweak of the individual commitments that would give fishermen privileges of recover relative to their commitments. This would truly be told to bring about the appropriation of an arrangement of tradable individual portions to manage admittance to the fishing of regular beds. The start-up of development on singular concessions faces the size of single amount charges applied by the organization for the development of clams and mussels. Would-be fish farmers consider that the scale in power is excessively high for firing up another arrangement of development where the degree of hazard is still high.

Hence the privatization of the stock gives off an impression of being a deciding element for the interaction of strengthening: from fishing to serious development, individualization of misuse (conventional gatherings, sport fishing affiliations, business armada proprietors, private farms), specialized increase (preservation, fishing, extensive aquaculture, intensive aquaculture), privatization of the stock (public, group, and individual possession), and section to the money related circle (protection, sport fishing, business fishing, aquaculture) all go inseparably (Asche et al., 2013).

In primary production where the assignment of the components is administered by the intricate situation of private possession, the rights expressly characterize the factors, the holders, the amounts, the offences and the approvals. They are elite as in every one of the additions and losses innate in the activity of a right tumble to the holders of the right. They are combinable and thus tradable. They are relevant and successfully applied since a right that isn't applied is certainly not a right (Turcios & Papenbrock, 2014).

The portability of vagile species makes it difficult to just translate to extensive development the arrangement received in agribusiness and shellfish culture – for example, the development of the srate - for the allotment of the stock. Nonetheless, a superior incident between the dispersion of expenses and advantages might be looked for by dispensing select quantitative privileges of alevin raising and recover to producers. Anyway, the utilization of this framework in development where similar species is fished in similar regions faces an issue: the difficulty of recognizing the developed or wild beginning of the catches. This vulnerability infers that fishing and development ought to be dependent upon a similar arrangement of guidelines. Yet, it is in no way, shape or forms sure that the increment underway anticipated from another arrangement of development will get the job done to persuade fishermen, who have open and free admittance to the wild stocks, that it is to their greatest advantage to embrace an arrangement of quantitative individual rights.

The appropriation of individual rights additionally brings up the issue of the decision of the allotment component. With the opening up of rustic social orders to public economies, the situation of standard rights dependent on friendly control is continuously falling into neglect. The terrible showing of conventional strategies for guideline of access in business fishing delineates the incapability of authoritative guidelines as a method of designating a factor of production in a business movement. The benefits of tradable rights (assurance of the cost of fishing rights by direct exchange between the gatherings) faces the resistance stimulated by the selection of this framework for distributing factors up until recently allotted by non-market instruments, and complaints of imbalance coming about because of individuals' various capacities concerning making the most of the new changes made by the appropriation of another arrangement of guideline.

The development of possession systems and instruments for exchanging rights, at last, relies upon the explanation of public wards which, adrift as ashore, characterize and ensure the proprietorship systems for regular assets. While the new law of the ocean does surely concede states privileges of semi responsibility for assets contained inside their select financial zones or exclusive economic zones(EEZs), their position stays weakened for the stocks whose srate of dissemination covers the EEZ and the high oceans and is shared for stocks whose region stretches out more than a few EEZs.

Figure 8.10: Evolution of Integrated Open Aquaculture Systems in Hungary: Results from a Case Study.

Source: file:///C:/Users/paulm/Downloads/sustainability-10-00177-v2.pdf

In any case, for a couple of exemptions - that of Iceland specifically, which is recognized by its topographical confinement and the drives that this nation has taken to change the association of its fisheries to new conditions -, the current circumstance in Europe barely appears to be great for the augmentation of extensive aquaculture to vagile species. The incredible improvement of mainland waters for the second decreases the possibilities for exceptionally transitory species like salmon (Van Doan et al., 2020). In most European fisheries, admittance to the assets stays generally open and free. In the EU, the Common Fisheries Policy has so far focused closer on the upkeep of business as usual and the minimization of the quick ramifications for fishermen of the intermittent measures taken for the preservation of stocks, than to the explanation of the system of responsibility for assets and the modernization of the components for controlling access.

In essential production where the assignment of the elements is represented by the intricate situation of private possession, the rights unequivocally characterize the factors, the holders, the amounts, the offences and the authorizations. They are elite as in every one of the increases and losses intrinsic in the activity of a right tumble to the holders of the right. They are combinable and thus tradable. They are appropriate and adequately applied since a right that isn't applied is anything but a right.

The portability of vagile species makes it difficult to just render to extensive development the arrangement received in agribusiness and shellfish culture – for example, the development of the srate - for the appointment of the stock. In any case, a superior fortuitous event between the circulation of expenses and advantages might be looked for by allotting select quantitative privileges of alevin raising and recovery to producers. Anyway, the use of this framework in development where similar species is fished in similar regions faces an issue: the difficulty of recognizing the developed or wild beginning of the catches. This vulnerability suggests that fishing and development ought to be dependent upon a similar arrangement of guidelines. However, it is in no way, shape or forms sure that the increment underway anticipated from another arrangement of development will get the job done to persuade fishermen, who have open and free admittance to the wild stocks, that it is to their greatest advantage to receive an arrangement of quantitative individual rights.

The reception of individual rights additionally brings up the issue of the decision of the assignment component. With the opening up of country social orders to public economies, the situation of standard rights dependent on friendly control is the step by step falling into neglect. The lackluster showing of traditional techniques for guideline of access in business fishing delineates the inadequacy of regulatory guidelines as a method of assigning a factor of production in business action (Verdegem & Bosma, 2009). The benefits of tradable rights (assurance of the cost of fishing rights by direct arrangement between the gatherings) faces the resistance stimulated by the selection of this system for dispensing factors up until recently designated by non-market components, and protests of imbalance coming about because of individuals' various capacities with regards to making the most of the new changes made by the reception of another arrangement of guideline.

The development of proprietorship systems and instruments for exchanging rights, at last, relies upon the explanation of public locales that, adrift as ashore, characterize and ensure the possession systems for normal assets. While the new law of the ocean does in reality concede states privileges of semi responsibility for assets contained inside their select financial zones (EEZs), their position stays weakened for the stocks whose rate of dissemination covers the EEZ and the high oceans and is shared for stocks whose region reaches out more than a few EEZs.

However, for a couple of exemptions - that of Iceland specifically, which is recognized by its geological detachment and the drives that this nation has taken to change the association of its fisheries to new conditions -, the current circumstance in Europe hardly appears to be ideal for the expansion of extensive aquaculture to vagile species. The extraordinary advancement of mainland waters for the second lessens the possibilities for exceptionally transient species like salmon. In most European fisheries, admittance to the assets stays generally open and free. In the EU, the Common Fisheries Policy has so far focused harder on the support of the state of affairs and the minimization of the prompt ramifications for fishermen of the intermittent measures taken for the protection of stocks, than to the explanation of the system of responsibility for assets and the modernization of the instruments for controlling access.

Chapter 9

Main Drivers of the Increase in Fisheries and Aquaculture Production

CONTENTS

9.1 INTRODUCTION

In a significant number of states around the world, the interest in fish has kept on developing. Without precedent for history, more fish for human utilization have started from farms than from wild catch, reaching nearly equality in 2012 as indicated by the most recent worldwide report from the Food and Agriculture Organization of the United Nations (FAO). Given population development, growing urbanization, and rising salaries in the creating scene, this pattern is relied upon to proceed. Since levels of catch fish production have deteriorated over the many previous years, the world will subsequently be more reliant upon aquaculture in the coming many years. After that, the fishery and aquaculture area is ending up at an essential junction, and when it is confronting quick mechanical change, expanding requests, and rising feed costs (Asche et al., 2013). The significance of fish and fishery-based exercises to food security in less-developed countries is exceptionally unmistakable in those networks taking part in little to medium-scale tasks in Africa and Asia. This is the consequence of both the utilization of fish in the families occupied with fishing activities just as the pay that these families create. It has been noticed that inside populations connected vigorously in fishing exercises, the fish from small-scale fisheries address an entire segment of animal protein allow and give a wide cluster of fundamental supplements. The requirement of fish from utilization by producers is frequently disparaged in worldwide measurements, resulting in a general disregard for the catch levels that come from small-scale fisheries—particularly those that are inland.

A report from an undeniable level board of specialists shows that, given that more than 80% of worldwide aquaculture production is delivered by little to medium-scale ventures (that are intensely gathered in Asia), the significance of fisheries to food security applies similarly to small-scale aquaculture tasks and small-scale catch fisheries. Numerous examinations have shown that the food security of families occupied with such aquaculture undertakings is improved through the money these activities create, just as from the expanded accessibility of fish items for their utilization. This has been noticed methodically in Bangladesh and India cases, just as in places in Africa south of the Sahara (like Malawi), where nearby networks have taken up small-scale aquaculture.

Evaluations distributed by FAO in 2014 hold that the portion of world wild stocks that are overexploited is on the ascent—with almost 29% of fish stocks misused at organically impractical levels in 2011.

Given the significance of animal-based proteins to supplementing human well-being (particularly the sorts of micronutrients found in fish-based proteins, similar to omega-3 unsaturated fats), fish contains a fundamental segment of a food-secure future. This part accordingly centers around the powerful job that aquaculture keeps on playing in giving animal proteins in human weight control plans around the world—and particularly in quick arising areas of Asia. Considering both the cutoff points to extending rangelands for animals and the environmental imperatives to expanding catch fishery production, aquaculture addresses the following and maybe even the final wilderness of large-scale animal protein production.

9.2 KEY DEVELOPMENTS FROM 2014

While there is critical vulnerability and discussion around a few issues in the fishery and aquaculture field, there is concurrence on numerous significant continuous patterns. The over exploitation of catch fisheries is apparent: estimates distributed by FAO in 2014 hold that the portion of world wild stocks that are overexploited is on the ascent—with almost 29% of fish stocks used naturally impractical levels in 2011.

Different procedures and ways to deal with assessment paint a considerably more critical picture. The record-high catch reaps in 2013 in any case; the more extended term viewpoint is clear: the complete worldwide catch production has leveled, staying stale for as far back as twenty years. Furthermore, a large number of the world's fisheries progressively comprise lower-value fish, demonstrating possible further consumption of the stock. Again, the stale reap comes at both a steadily expanding effort and cost (mainly fuel). In this manner, it is impossible that catch fisheries will want to give more fish essentially in the future than they do now.

The fast development of aquaculture conveys with it some significant trade-offs and limitations. Notwithstanding quick enhancements in taking care of innovation, the area requires developing volumes of feed. It is relied upon to keep swarming out different sites for fishmeal and fish oil, and it is additionally expected to require a lot of vegetable-based dinners, principally soy-based, even though there is progressing research on different sources, for example, dried distillers grains and solubles. As world costs of fishmeal stay at phenomenal highs, the aquaculture area keeps looking for approaches to utilize fishmeal and fish oil more as essential added substances during explicit phases of development instead of using them as a selective feed source (Troell et al., 2014).

While introducing an extraordinary opportunity, aquaculture improvement should be directed with high respect for the manageability of the strengthening of aquaculture. Endeavors should be made to misuse collective energies and to incorporate aquaculture with existing homestead exercises and harvests while depending on local fish species at every possible opportunity. The species' prudent decision should lower reliance on (typically imported) feed, compost, and anti-infection agents; lower hazard of illness and natural debasement; and further developed the flexibility to disease and environmental change. El Niño is required to happen before the year's over 2014 and last through spring 2015. Even though it is relied upon to be a frail occasion this time around, repercussions of such shocks on marine natural systems are not irrelevant, and the impacts on both marine-based catch and aquaculture are felt universally.

9.3 CONCESSIONS TO SUSTAINABLE SEAFOOD

Further developed administration and oversight of the worldwide fishing area is fundamental and is an issue with moderately broad understanding across critical partners. The reality of the overexploitation of wild fisheries is very much exhibited by the multiplication of worldwide drives and different endeavors to address them at all of the significant global political and advancement associations. A couple is featured hereby:

The United Nations (UN) met at two working gathering gatherings in 2014 to examine the chance of adding more instruments securing biodiversity to the UN Convention of the Law of the Sea (UNCLOS).

FAO has set up a Blue Growth drive as a reliable way to deal with the feasible, incorporated, and financially delicate administration of the seas.

The Organization for Economic Co-activity and Development (OECD) advances its own Green Growth Strategy.

The World Bank's Global Partnership for Oceans has coordinated a spearheading blue stripboard and, as of late, distributed a report that offers direction on how best to empower and spread the take-up of manageable practices in the fisheries value chain. The board addresses a huge discourse between scholastic specialists, significant arrangement organizations like the World Bank, and vital private-area delegates.

On January 1, 2014, the European Union started executing its new Common Fisheries Policy, which plans to eventually kill disposes of and set catch production at the most extreme manageable yield to preserve stocks.

On the off chance that the endeavors recorded above end up being fruitful, the advantages of further developed administration could be critical. FAO assesses that modifying wild stocks would bring about an extra worldwide catch worth $32 billion yearly.

9.4 POSSIBILITIES FOR FISH AND AQUACULTURE IN EMERGING ECONOMIES

Worldwide Prospects to 2030: Implications for Nutrition

In the as of late distributed worldwide standpoint report qualified to fish for 2030, the staggering more significant part of the vast expansion in worldwide fish production is relied upon to come from aquaculture—as it has for as far back as a very long while. The table below shows the production gains from different world areas and the offer that comes from aquaculture:

TABLE 1 Projections of fish production to 2030 under baseline scenario

	2010–2030 increase in production (million tons)	2010–2030 increase in production (%)	Share of 2010–2030 increase coming from aquaculture (%)
Africa south of the Sahara	0.3	4	64
Middle East and North Africa	0.8	22	97
India	4.8	60	98
Other South Asia	2.4	32	82
Southeast Asia	7.9	38	97
Japan	(0.5)	-9	–
China	16.5	31	101
Other East Asia and Pacific	0.3	7	105
Latin America and Caribbean	2.1	11	94
North America	0.2	4	103
Europe and Central Asia	0.8	6	122
Rest of the world	0.0	1	60
Global total	35.7	24	100

Figure 9.1: Table 1 - Projections of fish production to 2030 under the baseline scenario.

Source: https://www.ifpri.org/sites/default/files/gfpr/2015/feature_3087.html

While little gains in catch aquaculture are achievable in a predetermined number of areas, it is extended that any critical measure of extra fish aquaculture universally will come from aquaculture. Around half of the vast expansion in aquaculture, and in this way all-out fish aquaculture, is projected to happen in China alone, while all of Asia consolidated will involve very nearly 90% of the development in worldwide fish aquaculture. In locales that the table above shows will offer over 100% to future aquaculture increment, aquaculture is more than counterbalancing real reductions in catch aquaculture. At any rate, in any area with critical extended expansions underway, the commitment of aquaculture is, for the most part, near 100%.

This depiction shows the solid job that arising areas in both Asia and Latin America play in adding to worldwide fish supply. A couple of crucial nations, similar to China, are at the focal point of this development. However, different areas, identical to Southeast and South Asia, additionally have a significant job to carry out. One of the critical components of China's achievement in keeping a handy aquaculture area is that it has held a public expansion framework for aquaculture that has proceeded with far and wide effort to producers through all-around prepared and qualified staff. In the more extensive Asia locale, there is also a solid organization of preparing programs that help improve the aquaculture area. These remember the Network of Aquaculture Centers for the Asia-Pacific (NACA), the Southeast Asian Fisheries Development Center (SEAFDEC), the Asian Institute of Technology (AIT), just as such CGIAR organizations as the WorldFish Center. The various fish preparing focuses and administration focuses on shrimp producers have helped the efficiency of that area in Bangladesh. In Indonesia, a different Ministry of Marine Affairs and Fisheries (MAF) in 2000 significantly advanced in fostering a solid sectoral improvement methodology. The methodology incorporated the formation of advancement zones for aquaculture. The strengthening of aquaculture could be upheld by improving private incubators, dispersion and promoting channels for fish and fingerlings, preparing better data systems, supporting item accreditation, and admittance to capital. These kinds of substantial strategy-driven advancements are behind the solid development underway that is reflected in these projections (Teletchea & Fontaine, 2014)

The most significant portion of fish aquaculture in India is probably going to stay in the classifications of carp aquaculture, which have moderately low feed powers contrasted and other aquaculture species like Pangasius (catfish) or tilapia.

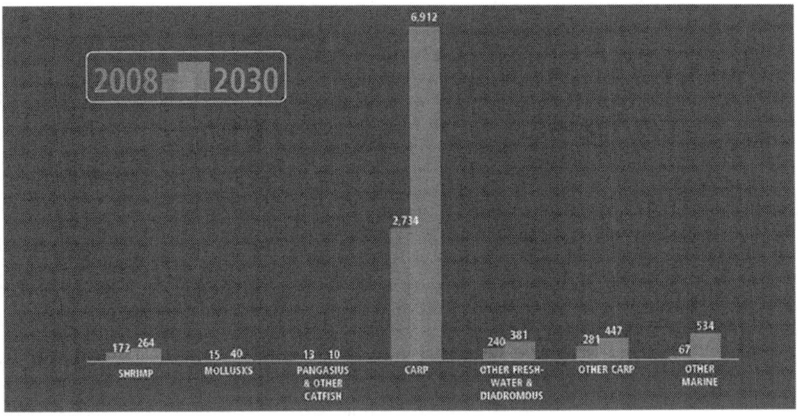

Figure 9.2: Projected aquaculture aquaculture in India by significant species groups ('000 tons).

Source: https://www.ifpri.org/sites/default/files/gfpr/2015/feature_3087.html

Regardless of their low energy and somewhat low value in exchange, carp stay a profoundly significant source of nourishment for fish-eating populations in South and East Asia and are essential for food security in those districts.

To get how this image affects nourishment, it is helpful to interpret the worldwide expansions in fish aquaculture and utilization from the fish to 2030 projections into changes in the per capita protein requirement, following across major territorial groupings of the world.

Figure 9.3 shows that the best-projected rate increments are in South Asia (particularly nations other than India, similar to Bangladesh) and China, with Southeast Asia expanding nearly as much in rate as North America. Africa and Latin America show a critical abatement in per capita requirement of protein from fish, given that their per capita utilization of fish is projected to diminish over the 2010–2030 period by 1.0 percent and 0.6 percent, individually, as far as yearly average development rates of per capita utilization . Currently, around 80% of worldwide fish aquaculture goes to meet human utilization needs, which doesn't change essentially over the projection skyline to 2030.

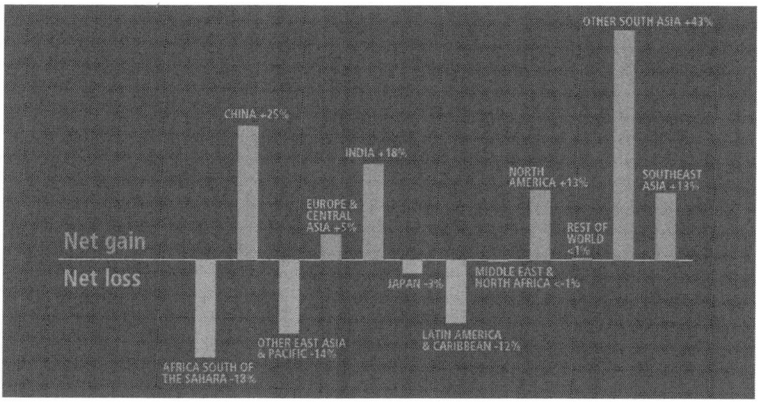

Figure 9.3: Projected change in per capita protein consumption from fish, 2010–2030.

Source: https://www.ifpri.org/sites/default/files/gfpr/2015/feature_3087.html

Besides contemplations of food security, a critical inquiry that remains parts concerning the expected blast of fish cultivating is the ecological impacts of such huge scope development stemming from inland freshwater aquaculture and farms in sensitive marine regions. Indeed, even regardless of the adverse natural effects of fish cultivating exercises, it is our view that aquaculture is a proficient method to source the essential animal protein and contrasts nicely with other animals systems. Aquaculture is likewise a nearly juvenile area countered and land-based domesticated animals tasks. Simultaneously there have been colossal upgrades inside the business as far as controlling disease while bringing down the utilization of anti-microbial, just as in further developing taking care of efficiencies across all types of aquaculture. Additionally, progressing work at CGIAR focuses on distinguishing procedures and advancements that can economically accomplish the fate of aquaculture imagined in this chapter.

9.5 PROCEEDING WITH GROWTH IN ASIA: RAPID SECTORAL CHANGE

One of the beautiful parts of worldwide fish markets in recent years is the marvelous development of aquaculture aquaculture in Asia, including the arising economies of South and Southeast Asia. Nations like Thailand and Vietnam are worldwide innovators in high-value aquaculture, providing a considerable portion of global aquaculture for shrimp, tilapia, and

Pangasius. By zeroing in on the global conveyance for the aquaculture-based aquaculture of Pangasius, the figure below exhibits the fast sectoral change arising in Asia.

Figure 9.4 likewise shows that nations that were not delivering any massive degrees of Pangasius in 1994 (like Vietnam) were at that point sizable producers five years after the fact and kept on developing to a point where they addressed 33% of the worldwide market by 2009.

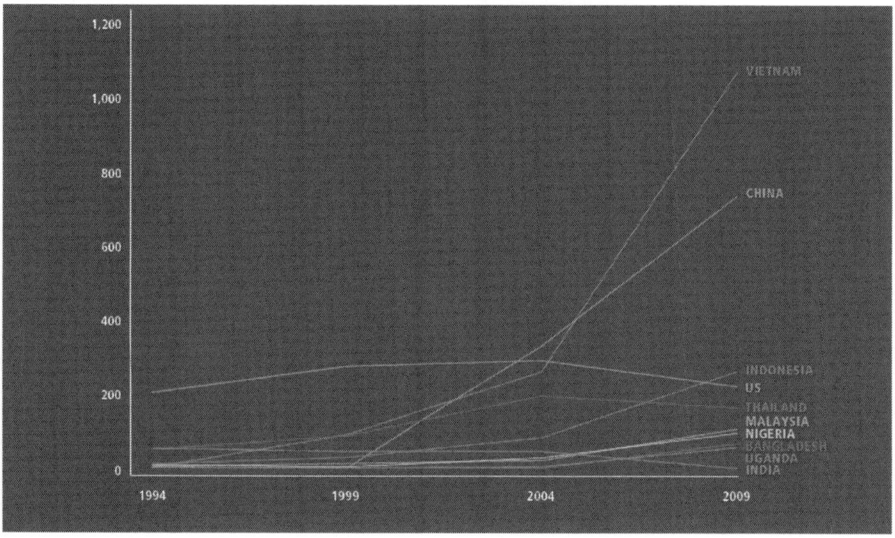

Figure 9.4: Major producers of Pangasius (catfish) from aquaculture ('000 tons).

Source: https://www.ifpri.org/sites/default/files/gfpr/2015/feature_3087.html

The substantial impact of Asia can likewise be seen on the exciting side. In the situations introduced in fish to 2030, the creators investigated the ramifications of proceeded (and surprisingly sped up) interest for fish in Asia, which addresses most of both utilization and aquaculture of fisheries items. The significant job of China in the worldwide equilibrium of organic fish market was noted in fish to 2020, a prior investigation by the International Food Policy Research Institute (IFPRI), and stays a point of convergence of premium in any examination of the worldwide fish economy. 16 Figure 4 shows the aquaculture projections for different fish species in China, utilizing a common situation under which China arrives at an all-out hydroponics aquaculture level across all types of only over 53 million tons by 2030. This level comprises around 57% of the worldwide hydroponics aquaculture in

2030, addressing an almost 46 percent expansion in hydroponics aquaculture over the 2010–2030 projection time frame.

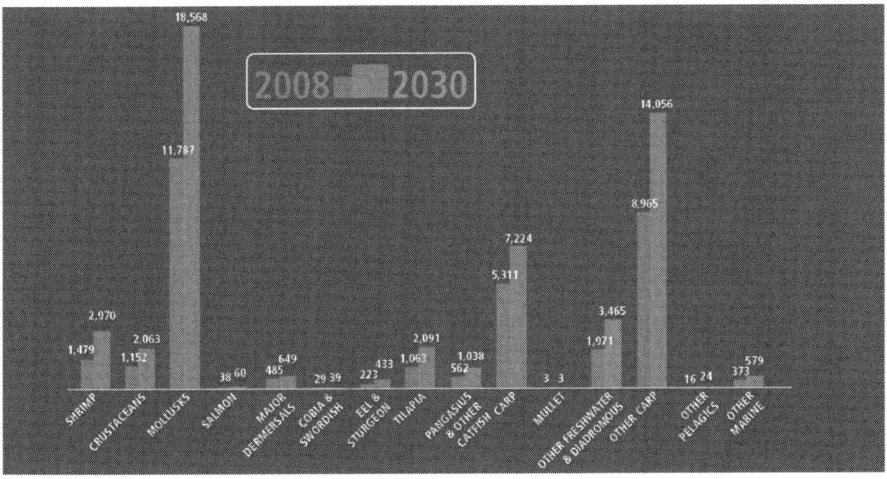

Figure 9.5: Projected hydroponics aquaculture in China by significant species groups ('000 tons).

Source: https://www.ifpri.org/sites/default/files/gfpr/2015/feature_3087.html

A fast change in customer inclinations, pay, or strategy in China would have immediate and wide-running impacts on the worldwide market of fishery items. The circumstance in China could (and likely will) without any assistance decide the fate of the whole worldwide market.

In the fish to 2030 report, a situation wherein a sped-up shift of purchaser inclinations toward higher-value fish was investigated to delineate the possible effect of China's developing customer economy on the general worldwide equilibrium of fish interest and supply. Under this situation, the per capita utilization of higher-value fish items like shrimp and different shellfish, just as salmon and fish, was projected to expand multiple times as quickly over the projection skyline to 2030, contrasted and the standard case. For other medium-value fishery items, like mollusks, the development in per capita request was multiplied comparative with the pattern level, while the inclinations for other fishery items continued as before (Sirakov et al., 2015).

Under this situation, the creators extended a general expansion in the worldwide fish supply of almost 23 million tons by 2030, with a large portion of the gradual aquaculture acquire coming from North America, Japan, and the remainder of East Asia and Pacific locale. The table below delineates the ramifications of this situation for utilization and exchange in China for different fish species.

	Food Demand		Net Trade
Shrimp	8,838	211%	–7,139
Crustaceans	3,079	205%	–1,250
Mollusks	18,506	105%	–12,513
Salmon	1,905	196%	–1,900
Freshwater and diadromous	783	3%	–615
Tuna	328	199%	–328
Other pelagics	19	-13%	–275
Major demersals	225	-4%	223
Other marine	63	-4%	59

Figure 9.6: Projected difference in food demand and net trade levels in China by significant species groups in 2030 under accelerated demand scenario ('000 tons and % change from baseline).

Source: https://www.ifpri.org/sites/default/files/gfpr/2015/feature_3087.html

The simulated situation of higher food interest in China brings about an extra interest in 2030, practically comparable to the complete anticipated worldwide expansion in fish production in the following 20 years. While the situation accepts critical development sought after in China, it shows that a quick change in purchaser inclinations, pay, or strategy in China would have immediate and wide-running impacts on the worldwide market of fishery items. The circumstance in China could (and likely will) without any assistance decide the fate of the whole worldwide market. Given the enormous offer that this locale has in the general worldwide interest for fisheries items, further investigations of financially and demographically determined interest development in Asia on territorial and worldwide fish markets are justified.

9.6 THE NEEDS FOR INVESTMENT IN AFRICA: THE ROLE OF AQUACULTURE IN AGRICULTURAL INTENSIFICATION

Inside the creating scene, the locale probably has the best unexploited potential for aquaculture development in Africa. In the fish to 2030 report, growth in aquaculture production for Africa was projected to be exceptionally low (as was displayed in the table above). This is above all else because the development pattern has been tiny previously. It is conceivable that aquaculture could take off in pieces of Africa, as was found on account of nations like Vietnam. On the off chance that this in fact was to occur, critical degrees of speculations would be required, and such a way would just start to show itself after progressive long periods of supported development in the area. Be that as it may, it is hard for any model to anticipate and mimic quick primary change prompting a huge departure (as was found in Asia).

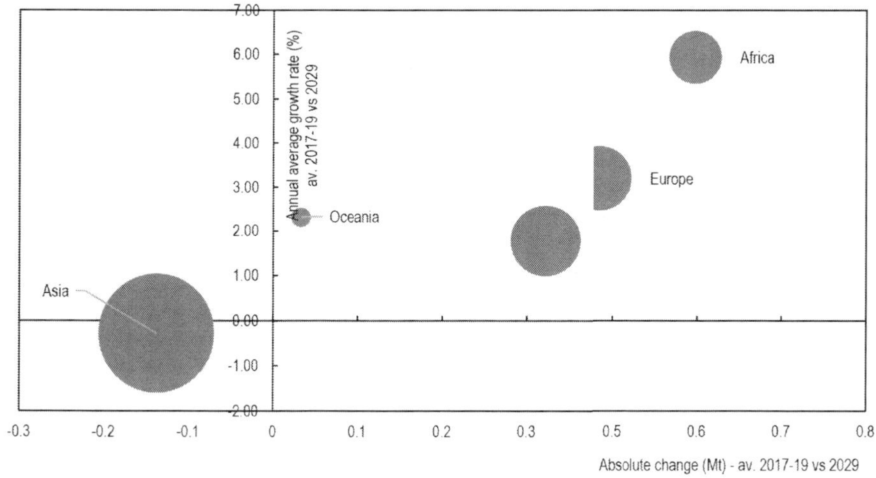

Figure 9.7: Growth in world capture fisheries production by continent.

Source: https://www.oecd-ilibrary.org/sites/4dd9b3d0-en/index.html?itemId=/content/component/4dd9b3d0-en#section-d1e19828

The colossal potential for such a direction exists; however it is coupled, without a doubt, with huge—yet not impossible—challenges. Besides the fundamental development underway itself, adequate consideration should be paid to satisfying the quality guidelines needed for accomplishing the development in trades expected to drive the future growth of the aquaculture area in Africa. Egypt is a genuine illustration of a nation that, despite offering solid strategy backing to its aquaculture area, saw its improvement of prices of ocean bream and ocean bass altogether impeded by its powerlessness to adjust to the European market's phytosanitary guidelines.

While a few nations may confront requirements in embracing more capital-serious sorts of aquaculture, there are places, for example, in Africa south of the Sahara, home to aquaculture much lower in capital force. Particular kinds of aquaculture, similar to tilapia, can be low in capital force and generally available for beginning areas to create, contrasted and higher-power species, similar to salmon. Under particular conditions and with the proper selection of species, aquaculture can turn into the most effective method of creating animal protein. Several vital types of fish (counting carp) can be delivered with insignificant data sources, while even species that utilize modest amounts of fishmeal can yield critical protein and significant unsaturated fats for human sustenance. This way, aquaculture items can be highly nutritious and reasonable, particularly in blend with minor local species that can be eaten entire, for example, the dagaa types of the Lake Victoria area.

That aquaculture improvement and extension are well inside the limit concerning agricultural nations is shown by the past accomplishment of aquaculture in Southeast Asia, South Asia, and Central America. Given certain speculations and an empowering strategy climate, a comparative example of overcoming adversity ought to likewise be conceivable in Africa south of the Sahara—with key nations like Ghana, Kenya, and South Africa being potential innovators in that pattern.

A vital factor for progress is the determination of the suitable species to develop. This incorporates contemplations of the ideal blend of feed force, yield, disease resistance, and simplicity of getting fingerling stock. Given the moderately high transportation costs that African producers will keep on confronting comparative with their Asian rivals, almost certainly, the aquaculture area in Africa will zero in on high-value species and those with high value added to the economy.

To help this objective, policymakers need to hasten the progression of outside capital into the area. This is vital for building up substantial scope activities with the fitting advancements for feed and disease. There is additionally the requirement for a committed aquaculture agency or service to assist with planning speculations, money, and guidelines with different pieces of public strategy. The adequacy of this methodology has been exhibited in

Madagascar, which has been effective in drawing financial backers to its shrimp area. South Africa's obligation to supporting aquaculture is another guide to follow: it set up an aquaculture park to draw in unfamiliar financial backers, a methodology displayed on the devoted aquaculture zones set up in Asia to think speculation and work with specialized effort and take-up (Stenton-Dozey et al., 2021).

That aquaculture advancement and extension are well inside the limit regarding agricultural nations is shown by the past accomplishment of aquaculture in Southeast Asia, South Asia, and Central America.

Besides tending to the absence of interest in the aquaculture area, there is likewise a solid requirement for strategy and administrative changes in the seas that harbor Africa's catch fisheries—the locale's primary fish supply. The infringement of alien armadas into the marine waters of African nations represses the maximum capacity of catch fisheries in the area. This is because these fishing zones experience the very open-access and unregulated overexploitation of fish stocks that tormented a large number of the world's fisheries before the adjustment that tradable amount plans and other arrangement measures brought to the Atlantic and Pacific waters. A few investigations recognize African coast waters as the locales with probably the most significant reductions in getting, noticing the vast vulnerabilities over the actual condition of the fish stocks.

Improving a further developed and productive fish cultivating area can halfway balance these issues in the future by giving an elective cause of supply to sea catch fish. This would assist with diminishing the cost of fish for customers while likewise decreasing the motivators for sea fishing armadas to overexploit populations of fish, consequently allowing them the opportunity to reconstruct their stocks and recapture their bioecological balance. Maybe then being viewed as a contender to the conventional catch fisheries area, aquaculture ought to be embraced as a corresponding action with enormous potential cooperative energies.

9.7 TRADING FISH: TENSIONS AND TRADE-OFFS BETWEEN EXPORT VALUE AND LOCAL NUTRITION

The substantial interest for fish items from arising nations is relied upon to keep developing, yet the physical requirements of sea-based environments imply that the conventional catch fisheries area can not build its stock past authentic levels. The aquaculture area will subsequently have to heighten its production reasonably.

A few issues identified with exchange fish merit a unique notice. Fish are intensely exchanged, a pattern that will proceed into what's to come. The fish exchange is essential for created and agricultural nations the same. As indicated by FAO's The State of World Fisheries and Aquaculture 2014 report, fish sends out top the rundown of non-industrial nation farming prices by dollar value. Without insecurity, a chance exists for non-industrial nations to gain by the fish exchange. However, this chance intrinsically conveys a trade-off between send-out value and neighborhood sustenance. These trade-offs are much the same as those in other farming areas, and they are firmly identified with the pressures among huge and small-scale fisheries.

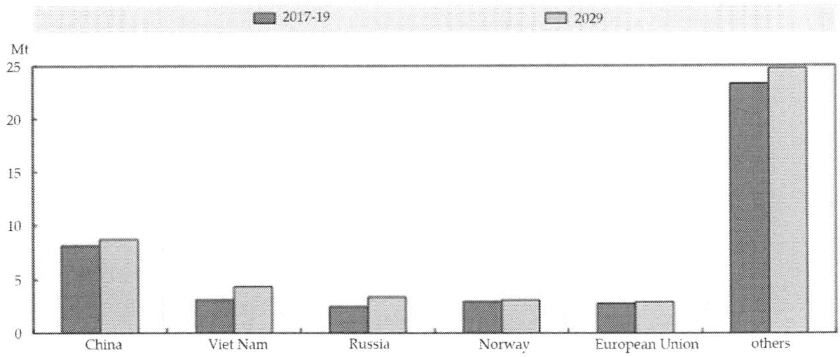

Figure 9.8: Exports of fish for human consumption by significant exporters in 2017-19 and 2029.

Source: https://www.oecd-ilibrary.org/sites/4dd9b3d0-en/index.html?itemId=/content/component/4dd9b3d0-en#section-d1e19828

Certain aquaculture items, mainly the higher-value ones like salmon and shrimp, are dominatingly bound for prices to higher-pay nations, where a more significant, generally urban, working-class exists. As such, these prices can give a considerable stream of price income. The production of a portion of these higher-value species will generally be more capital intensive, which assumes an undeniable degree of unfamiliar venture and innovation move, just as economies of scale. Meat-eating species especially require more feed, particularly fishmeal, and are in this way more intense. Such sources of information may likewise be imported. If the fishmeal is sourced locally—by squashing nearby minor species that would somehow or another have been consumed straightforwardly—a trade-off exists between send-out value and neighborhood sustenance (Sorgeloos, 2013).

Hence, legislatures of non-industrial nations should be mindful of these struggles when drafting their aquaculture improvement programs and techniques. Sometimes it very well may be fitting to zero in on more minor capital-serious species, particularly those with a more vegetable-based eating regimen. Such a methodology would mitigate a portion of the worries that have been communicated about the "food versus feed" rivalry between the lower-value species that would make some way or another go-to human sustenance in more unfortunate families, from one perspective, and their utilization in making fishmeal to take care of higher-value species for utilization by higher-pay purchasers, on the other.

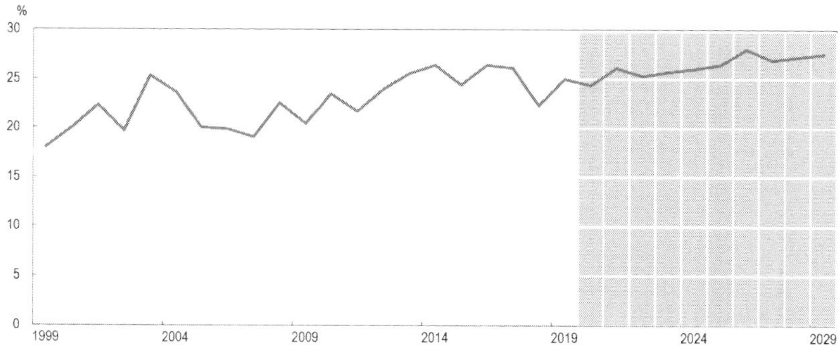

Figure 9.9: Share of total fishmeal production produced from the fish residue.

Source: https://www.oecd-ilibrary.org/sites/4dd9b3d0-en/index.html?itemId=/content/component/4dd9b3d0-en#section-d1e19828

9.8 MARKET SCENARIO

After solid development in 2018, with general production, exchange, and utilization coming to notevaluey pinnacles, the worldwide fisheries and aquaculture area declined marginally in 2019.1 Aquaculture production kept on growing by more than 2%, while catch fisheries declined by about 4% because of lower profits of specific species including cephalopods, cod and chose minor pelagic species.

By and large, 2019 contrasted with the earlier year. This was essentially because of value decays for some critical cultivated species, including shrimp, salmon, Pangasius, and tilapia, yet in addition for canned fish as a result of supply dominating interest. Monetary compression in certain nations, alongside exchange pressures between some chose key producers and shippers, added to a slight decrease in the worldwide exchange of fish and fish results of about 1% in 2019 contrasted with 2018 in both volume and value.

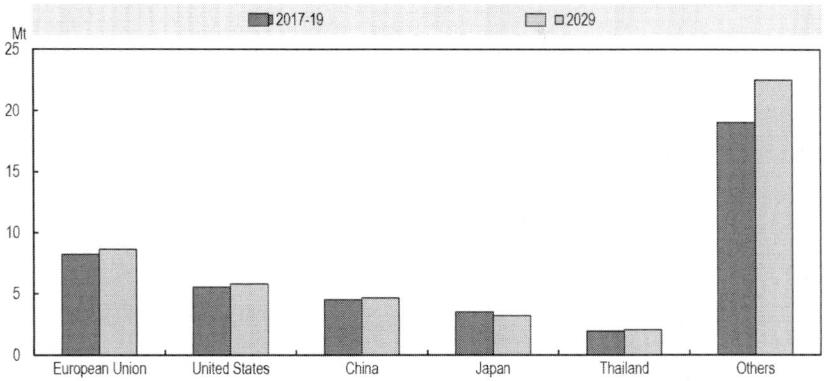

Figure 9.10: Imports of fish for human consumption by significant importers in 2017-19 and 2029.

Source: https://www.oecd-ilibrary.org/sites/4dd9b3d0-en/index.html?itemId=/content/component/4dd9b3d0-en#section-d1e19828

9.9 PROJECTION FEATURES

Comparative with the base time frame (2017-2019 standard), declared fish costs will increment at a rate between 1.5-2.1% p.a. over the following decade. In simple terms, fish costs are projected to remain generally unaltered regularly over the 2020-29 period – marginally negative at fish oil prices, exchanged fish, catch fisheries, and fishmeal; and somewhat sure for aquaculture species (Skerritt et al., 2020). Regardless of contrasts in the extent, all expenses are required to pursue close directions, with minor expansions in the central portion of the standpoint time frame, trailed by a decrease in the subsequent half. The variables adding to that decrease remember quicker development for Chinese production and a diminished tension on fish interest because of the recuperation of the pork area (as fish is one of the substitutions for pork utilization) from the significant episode of African swine fever (ASF). Because of the supported interest for fishmeal and somewhat stable inventory, the cost of fishmeal will keep on expanding marginally comparative with oilseed dinners.

Worldwide fish production is projected to arrive at 200 Million tons by 2029, expanding by 25 Million tons (or 14%) from the base time frame (average of 2017-19), however at more slow speed (1.3% p.a.) than more than the earlier decade (2.3% p.a.). This lull of development is driven by the collective impact of lower yearly development rates for both catch fisheries and aquaculture. One of the primary elements behind these regular lower rates is the presumption that China's fishing and aquaculture strategies for the following decade will line up with its thirteenth Five-Year Plan (2016-20), which moved needs towards advancing supportability and the modernization of the area with beginning limit decrease, to be trailed by a normal quicker development, specifically of aquaculture production. This is especially important as China is a wide margin the world's driving fisheries and aquaculture maker.

The commitment of aquaculture to worldwide fish production should keep on developing and outperform that of absolute catch fisheries (counting the sum used for non-food utilizes) by 2024. By 2029, aquaculture production is projected to arrive at 105 million tons, 10 million tons more than the catch area. Somewhat low feed costs are likewise behind the future development of aquaculture, and the benefit in the area is relied upon to stay high in the following decade, particularly for species that require modest quantities of fishmeal and fish oil. Catch fisheries are projected to see a moderate expansion underway more than the subsequent decade (0.4% p.a.), strikingly because assumptions that further developed administration in a few locales will keep on paying off, including through maintainable expanded usefulness of fish stocks. The portion of catch fisheries production changed into fishmeal, and fish oil will stay stable at about 18%. Nonetheless, all-out fishmeal and fish oil production are projected to increment by 10% and 17%, individually, over the course of the following decade, basically mirroring a more prominent utilization of fish buildups in their production. By 2029, the extent of complete fish oil got from fish waste is projected to develop from 41% to 45%, while for fishmeal, this extent will increment from 24% to 28% (Shaalan et al., 2018).

9.10 PRODUCTION

Worldwide fish production (from catch fisheries and aquaculture) is projected to increment from 176 Million tons in the base time frame to 200 Million tons by 2029. While this addresses an extra 25 million tons of fish each year by 2029, the rate of development and the full expansion underway keep on falling. In total terms, the expansion in all-out fish production over the viewpoint time frame is projected to be 73% of that seen over the earlier decade, when world fish production was 34 Million tons higher by the last year. Likewise, the rate of development (14.0% over the standpoint period) should be lower than the 23.8% accomplished over the earlier decade. This reflects lower development rates in the aquaculture (2.3% p.a.) and catch fisheries production (0.4% p.a.) when contrasted with the earlier decade (4.3% p.a. for aquaculture, 0.7% p.a. for catch). Regardless of this more slow growth, aquaculture will stay the principal driver of development in fish production at the world level. Developing from a portion of 47% of total fish production in the base year, aquaculture production is projected to overwhelm catch fisheries in 2024 and reach 52% by 2029.

Aquaculture production is projected to be 105 Million tons by 2029, an increment of 28.4% compared with the base time frame, contrasted with 59.6% in the earlier decade. This expected lull in aquaculture production development will fundamentally be brought about by lower usefulness gains, more stringent ecological guidelines, and expanding shortage of appropriate areas because of the contest from other land and water clients. China, the world's biggest aquaculture producer, is relied upon to encounter a considerable decrease in its cultivated fish production development rate. This outcome from the execution of new approaches preferring the supportability and modernization of the area, which are projected to result in more slow development rates toward starting the standpoint time frame before getting again towards the finish of the viewpoint time frame. Generally speaking, Chinese aquaculture production is projected to develop by 24.5% in the following decade, contrasted and 46.6% in the earlier decade. By 2029, China is relied upon to represent 56% of worldwide aquaculture production, determined and a portion of 58% in the base period. At the world level, a bit of the diminished development in Chinese production will be relieved by more significant expansions underway somewhere else. Aquaculture production is required to develop on all landmasses. Most world fish production will keep on starting from Asia, with a projected 89% of world aquaculture production by 2029 (Shah et al., 2018).

Development rates will differ across species, bringing about a difference in the synthesis of aquaculture production by 2029. The portion of dominant cultivated species, like carps and mollusks, is projected to decrease to 56% by 2029, down from a pinnacle of 77% during the 1990s. The portion of shrimps and prawns and tilapia and catfish (counting pangas) will increment over the viewpoint time frame.

Catch fisheries production is projected to increment somewhat more than the following decade (0.4% p.a.). By 2029, catch fisheries production is projected to arrive at 95 million tons, up 1.3 Million tons contrasted and the base period. This vertical pattern is because of assumptions for further developed profits in some fishing regions invigorated by the moderately high fish costs or where loads of specific species are recuperating, likewise connected with better administration which can expand the efficiency of fish stocks and consequently catch prospects, just as from diminished on-board waste and disposes of. The most elevated development rate and most significant expansion in volume terms in catch fisheries production are standard in Africa. At the same time, Asia is the solitary landmass projected to encounter a decay, basically connected to the average decreased catch fisheries production of China (- 10% by 2029 contrasted with the base time frame). On a nation premise, the most significant expansions in catch fisheries are projected in the Russian Federation (+0.6 Million tons), the Philippines (+0.3 Million tons), and Indonesia (+0.3 Million tons). In the long stretches of the El Niño6 wonder, catch fishery production in South America will decay, bringing about world catch production falling around 2% during these periods. The portion of catch fisheries production bound for direct human utilization is projected to increment from 77% in 2017-2019 to 79%, or 2.3 million tons, by 2029. The more significant part of the excess 20 Million tons is relied upon to be decreased to fishmeal and fish oil.

9.11 CONCLUSION

Through a mix of general level methodology setting and prioritization, private-area speculation, and multilateral help and backing, a solid and dynamic aquaculture area could start to arise in critical African nations and add to the substantial worldwide development effectively been happening in late many years.

This part has highlighted a portion of the critical difficulties of nations seeking to develop their fishery and aquaculture areas. The substantial interest for fish items from arising countries is required to keep creating, yet the biological limitations of marine environments imply that the traditional catch fisheries area can not build its stockpile past authentic levels. This way, to fulfill this need, the aquaculture area (as a natural substance of private makers) should reasonably escalate its aquaculture. Extra upgrades in feed and illness the control should emerge for this to occur (Wurts, 2000).

This is happening among developed and arising producers in Asia and Latin America because of the solid cost signals produced by expanding requests. By consistently receiving an assortment of best practices in their aquaculture activities, these producers are contending effectively with their partners in North America and Europe.

The viewpoint is less sure for Africa. Later on, the district's extremely low aquaculture aquaculture may invert itself if Africa can imitate the quick scale-up that has been found in Southeast Asia and Latin America. Be that as it may, this will require the private area to give the necessary aptitude and innovation acquired from these different areas. Such a change will possibly come to fruition if strategy-driven motivators are given to the private srate in a manner that sets up devoted zones for aquaculture, combined with certain rights and admittance to capital and specialized help. The solid expert organizations that help aquaculture aquaculture in Asia (like NACA and SEAFDEC, as noted prior) are additionally gravely required in Africa south of the Sahara. Such organizations can work with the specialized preparation of producers and speed up the take-up of efficiency upgrading practices and advancements. The job of global focuses, including the WorldFish Center, in working with the spread of information and best practices is likewise significant in supporting the development of the aquaculture area in Africa. So too is the extra help that could emerge from bodies like the World Bank and the International Finance Corporation (Yang et al., 2020). Aside from offering direct service to the private area through advances and credit offices, these multilateral establishments can likewise give African nations immediate specialized help and direction in working on the administration of the fisheries area and in the plan of public-area foundations. In aggregate, through a blend of general level system setting and prioritization, private-area speculation, and multilateral help and backing, a solid and lively aquaculture area could start to arise in vital African nations and add to the substantial worldwide development that has effectively been happening in ongoing a long time in different areas of the production scene.

Chapter 10

The Future of Aquaculture Fish Farming

CONTENTS

10.1 INTRODUCTION

Over the most recent 35 years, marine aquaculture production has developed from 1,000,000 tons each year to around 55 million tons. Development proceeds at about 10% each year, and for specific species, the amount delivered by aquaculture surpasses the amount fished. Notwithstanding, we realize that there are bottlenecks for additional advancement of marine aquaculture and, in any event, for its endurance, essentially as polished today. For instance, sooner rather than later, it will be important to evaluate the energy equilibrium of aquaculture: the production of predator species isn't energy proficient, and energy losses at each progression in the production chain could be kept away from if the essential material of feedstocks (anchovies, sardines, mackerel) was consumed straight by people and aquaculture focused more on herbivorous and detritivorous species.

Figure 10.1: The future of aquaculture, two sides of the story.

Source: https://blueocean.net/future-aquaculture-two-sides-story/

The cultivating of fish in land-based and seaward fish farms has two basic angles: the accessibility of protein to create modern feedstock and the effect of fish-farm wastewater on marine coast regions. The most basic part of mollusk cultivating is intensive plants and the strategies used to collect from normal populations. There are also numerous different issues related to aquaculture that might be no less basic than energy utilization and natural harm (Yang et al., 2020). They incorporate the presentation of allochthonous species, which might be vectors of viral and bacterial contaminations, major or even misrepresented utilization of pesticides and anti-infection agents (particularly in the new past), and change of benthic and tiny fish networks. To handle these issues, the FAO prescribes an environment way to deal with aquaculture, coordinating "the movement inside the more aquaculture environment so that it advances practical turn of events, value, and flexibility of interlinked social and natural systems."

Figure 10.2: Norway produces 33% of the world's farm-raised Atlantic salmon, with Chile producing 31%, followed by Scotland and Canada.

Source: https://blueocean.net/future-aquaculture-two-sides-story/

10.2 ACCESSIBILITY OF PROTEINS

The protein necessities of most industrially raised fish are high, going from 30-55% of their eating regimen. Modern feed is utilized to deliver around 29 million tons of fish and prawns and uses enormous amounts of fish oil and fishmeal produced using wild "bluefish" (anchovies, sardines, mackerel), particularly for items focused on business sectors of rich nations. Significant production of fishmeal happens in South America, Asia-Oceania, and Europe. The assessment directed in the North Atlantic, a region characterized as being in danger of pollution by the European Food Safety Authority (EFSA), shows that aquaculture systems regularly favor the production of fish that are more contaminated than wild species, as on account of European cultivated salmon, and that the level of pollution is frequently because of the utilization of profoundly contaminated feedstocks.

European fishmeal, predominantly created in Denmark, Norway, and Iceland, is among the most elevated foreign substances. These collect more in greasy species, like salmon and eels. It is likewise realized that cultivated fish are more extravagant in fat than wild fish. Proteins and oils of plant beginning are accessible in huge volumes (Zhang et al., 2015). Examination shows that many of these essential materials are appropriate for use in feedstocks and can somewhat supplant feedstocks containing fish oil and fishmeal. Further advancement of aquaculture should in this manner include a decrease in the amount of fish oil and fishmeal in feedstocks because distinctive healthful administration of cultivated fish could diminish the danger of industrial or natural toxins and because substitution of these feed segments would lessen the effect of overfishing on wild fish stocks.

Figure 10.3: The United States imports $13 billion of seafood annually.

Source: https://blueocean.net/future-aquaculture-two-sides-story/

Momentum-adjusted feedstocks don't appear to address their issues completely: they support that fish development isn't something similar and that the fish are more vulnerable to infection. More examination is required to build plant proteins, and it is important to conquer the resistance of many bream and ocean bass farmers. There is, in any case, the chance of utilizing fish dinner and oil acquired from side-effects of the handling of cultivated and wild fish if it has not been cooked at temperatures above 90°C which denatures the proteins.

The biofuels business is creating protein side-effects that have not yet been completely assessed. For this situation, impurities aren't settled; however, overfishing of wild stock is moderated. Past this, green growth and bio proteins got from microorganisms can turn out to be financially cutthroat, additionally giving long-chain omega-3 unsaturated fats (Sandersen & Kvalvik, 2015). The AQUAMAX project (practical aquafeeds to boost the medical advantages of cultivated fish for consumers), financed by the EU and facilitated by the National Institute of Nutrition and Seafood Research (NIFES) of Norway grouped assessment groups from numerous European and extra-European nations. The point of the task was to supplant fish dinner and oil in feedstocks for salmon, trout, carp, ocean bass, and bream however much as could be expected with practical contaminant-free alternatives (foreign substances like dioxins, PCBs, and fire retardants) while keeping up with the medical advantages of fish utilization.

Figure 10.4: Atlantic salmon caught in Puget Sound, Dean Rutz.

Source: https://blueocean.net/future-aquaculture-two-sides-story/

A goal was to decrease fishmeal to 12-16% and fish oil to 8-12% for salmon, 5% and 5% separately for trout, 15% and 10% individually for ocean bass and bream. The elective feeds appeared to keep up with tissue quality, including omega-3 substances. The last evaluations were directed with youngsters and pregnant ladies, in whom a critical take-up of omega-3 fats was noticed. Eutrophication According to experts, intensive aquaculture can have ecological effects inshore and seaward. The arrival of phosphorus is 19.6-22.4 kg/ton of item, 34-41% of which is delivered in a broken-down structure, while dispersal of nitrogen is 52-95% of the nitrogen in the feed. In a Mediterranean land-based fish farms producing bream and ocean bass, over 80% of the broken down inorganic nitrogen in wastewater was ammonium, and 86.2% of the inconstancy of complete smelling salts was feed-subordinate, showing the significance of right taking care of. Intense land-based fish farms produce particulate high in the natural matter (feed deposits, dung, algal debris, and so forth), released by wastewater and scatters in the last receptor environment, causing water turbidity encompassing marine regions and advancement of silt with natural matter (Rosas et al., 2019). This progression of natural carbon can essentially agitated residue biogeochemical measures, change benthic networks, and make states of anoxia. Marine algae knolls close to fish farms may experience a big effect. Aquaculture has been the reason for the far-reaching loss of marine phanerogams (2% each year) all through the world. Decreasing effects on a base is a logical test, and numerous producers have packed their consideration in this field. Bivalves in land-based systems or close to marine plants, where they can develop more than in the vast ocean, could diminish influence.

Figure 10.5: Farmed salmon out-competed by wild salmon.

Source: https://blueocean.net/future-aquaculture-two-sides-story/

10.3 HARVESTING OF MOLLUSKS

Harvesting of wild mollusks additionally thought to be an aquacultural practice should likewise be referenced. Mollusks are progressively obtained by automated strategies that intensely affect the substrate. This is valid, for instance, for the collecting of *Solen marginatus*, *Tellina*, *Callista chione*, and different mollusks, which is led with turbojets that suction and sifter immense amounts of sand, and for reaping of the allochthonous *Tapes philippinarum* in the tidal pond of Venice, which is completed with an extraordinary assortment of gadgets. These activities influence a deep layer of silt for many hectares, annihilating the biocoenoses present, destroying plants, and defending their foundation. Measures to ensure the Mediterranean coasts are being examined at the European level to pass laws, actually questioned and deferred, restrict collecting and lessen influence.

10.4 FISH FOR 9 BILLION INDIVIDUALS

The worldwide population is developing at an amazing speed. In 1950 the world had an aggregate of 2.5 billion individuals, a figure that had reached 7 billion by 2012. As indicated by United Nations estimates, this number could surpass the 9 billion by mid-century. As populations increment, so too does the requirement for food. Fish is an inescapable, reasonable, and solid source of important protein. There is no doubt, hence, that the worldwide interest in fish will escalate in the future.

At the point when we consider that the measure of wild-caught fish has not expanded lately, just a single elective remaining parts: fish cultivating, or aquaculture, should fill the gap. Is it fit for doing as such? This is the issue numerous researchers throughout the planet are attempting to reply.

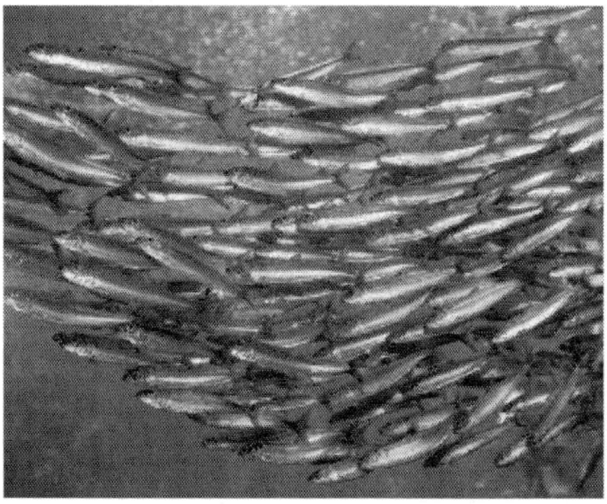

Figure 10.6: What Feeds Fish Farms?

Source: https://blueocean.net/future-aquaculture-two-sides-story/

For a long time, aquaculture assumed a somewhat minor part in worldwide fish production, yet its importance has expanded drastically in recent years, prodded by the interest from Asia's quickly developing populations. Today, aquaculture makes a significant commitment to human sustenance. For instance, it gives an enormous extent of the animal protein consumed in China, Bangladesh, and Indonesia. Worldwide production of fish, mussels, and crab in 2010 was right around 60 million tons, which remembers production for marine waters, saline water, and freshwater. Aquaculture production is currently around 3/4 of that from sea fish and fish trapped in nature. In 2011 this added up to 78.9 million tons. No other food industry has shown such development as aquaculture in ongoing many years. Somewhere in the range of 1970 and 2008, yearly production overall expanded by a normal of 8.4 percent, significantly more than poultry cultivating and egg production, which have the second most elevated development rates after aquaculture.

10.5 THE BENEFITS AND DRAWBACKS OF AQUACULTURE

Aquaculture has come in for some heavy analysis lately. For different reasons, it draws in contention. Food, fecal and metabolic waste from intensive fish homesteads can prompt the eutrophication (over-treatment) of water in waterways and marine bayous. There have likewise been protests that fish cultivated under intense conditions for the greatest yields are more powerless to infection than their family members in nature (Naylor et al., 2021). Colossal measures of anti-microbials and different prescriptions are utilized to battle illness, especially according to shrimp on farms in South East Asia – with unforeseeable ramifications for encompassing biological systems and purchaser wellbeing. Sometimes these focuses are substantial; however, they ought not to degrade how aquaculture can be an exceptionally proficient and practical technique for providing people with animal proteins – and checking over-fishing.

Figure 10.7: Speeding the growth of farmed fish.

Source: https://blueocean.net/future-aquaculture-two-sides-story/

The cultivating of the exemplary normal carp or mirror carp gives a positive illustration of ecologically strong aquaculture. Carp are bottom dwellers, eating little sea-going animals, plants, dead plant matter, and waste material that accumulate on the lake floor. Carp lakes regularly have extremely clear water. They additionally strainer the water to remove suspended solids, subsequently assisting with keeping the water clean. Intensive mussel cultivating likewise assists with keeping the water clean. Mussels channel much water, sieving out little particles of food, in this manner neutralizing the over-preparation of water and algal sprouts.

Albeit the supplement-rich effluents from aquaculture offices can prompt issues in waterways or coastal regions, in any case, many fish farms are more harmless to the ecosystem than, for example, the intense cultivating of pigs or cows.

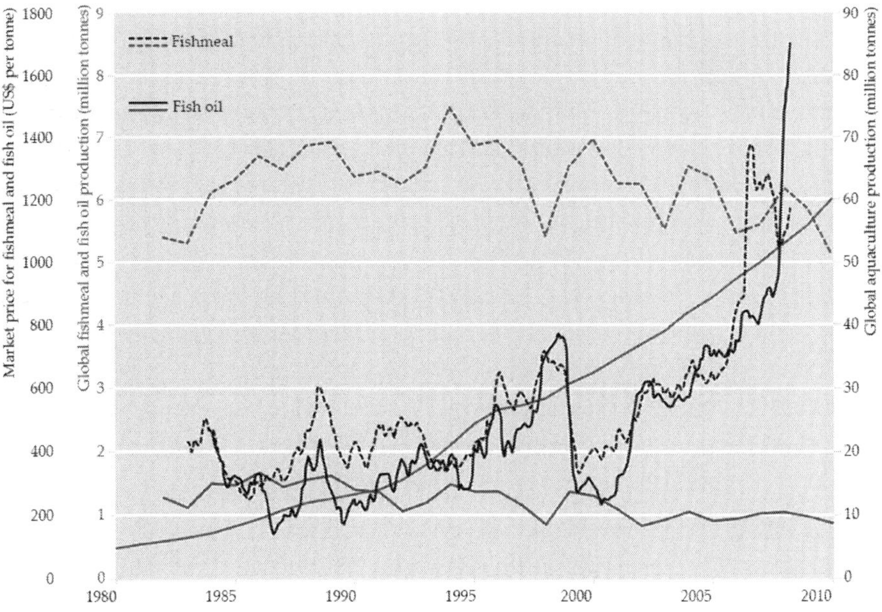

Figure 10.8: Although the output of aquaculture has increased greatly over the past 30 years, the sector's consumption of fishmeal and fish oil is at about the same level as in the 1980s. Greater use of plant-based nutrients is one reason; more efficient fishmeal and fish oils utilization are another. The price of fishmeal and fish oils has multiplied, mainly due to rising demand in China.

Source: https://worldoceanreview.com/en/wor-2/aquaculture/.eco-friendly-aquaculture/

The last discharge huge amounts of nitrogen and phosphorus from the slurry and fertilizer used to treat the land. Aquaculture produces far lower emanations of nitrogen and phosphorus and can generally measure up to those from the considerably less tricky cultivating of poultry. This is made crystal clear by the case of the Mekong Delta. Simply around 1 to 2 percent of supplement inflows into the delta come from Pangasius aquaculture. The greater part comes from farming, vegetables, and organic products, just as untreated city sewage and modern flows. Aquaculture additionally scores well when contrasted with domesticated animal rearing since fish and other sea-going natural entities need less sustenance to fabricate weight than land animals. Consequently, much less feed is needed to create 1 kilogram of carp than to deliver 1 kilogram of chicken, meat, or pork. One justification is that fish are cold-blooded animals, implying that their internal heat level is equivalent to that of their environmental factors.

They, in this manner, need less energy to create heat than warm-blooded well-evolved animals or birds. Additionally, it takes more prominent use of energy to continue ashore than in the water. As water is denser than air, it gives a light lift to the body, implying that fish are upheld without the advancement of hefty skeletal mass. Numerous marine animals like mussels, snails, and ocean cucumbers additionally oversee without an inner skeleton. This saves them the energy they would somehow or another utilization to synthesize bones. Fish have another energy advantage, as well: they are equipped for delivering into the water (as ammonium, a basic substance compound) any overflow nitrogen they may have retained with their food. Interestingly, land-based animals need to utilize energy to change over nitrogen into urea or uric acid. Just in this compound structure are they ready to discharge the nitrogen with their excrement or urine (Milstein, 2005).

In a worldwide coordinated effort, researchers have explored whether aquaculture and business fishing will be equipped for satisfying the worldwide need for fish in 2050. They are hopeful, accepting that indeed, they can. However, this would rely on the world's fish stocks being overseen reasonably in the long haul. Additionally, the fish utilized as feed-in aquaculture as fishmeal and fish oil should be used more productively. Researchers have additionally questioned the anticipated direct effect of environmental change and sea warming on potential marine fisheries production. They have reasoned that the measure of wild marine fish accessible for fisheries worldwide will likely be rearranged because of environmental change and anticipate a slight 6 percent by and large increment on marine fisheries potential.

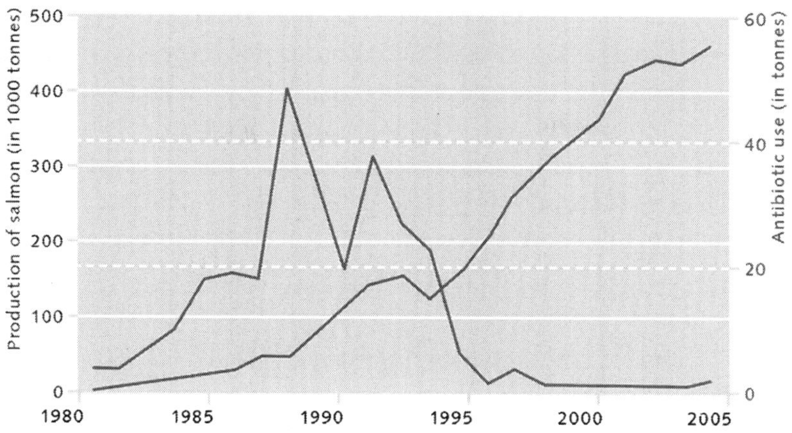

Figure 10.9: The example of Norway shows that the intensification and professionalization of production can lead to improvements. Despite increasing numbers of salmon, the use of antibiotics in the Norwegian salmon farming industry has declined.

Source: https://worldoceanreview.com/en/wor-2/aquaculture/.eco-friendly-aquaculture/

10.6 THE EVENTUAL FATE OF CULTIVATED FISH

Driven mostly by enormous population development, urbanization, and expanding abundance in Asia, aquaculture has developed by a decent 8 percent for each annum in recent years – quicker than some other food areas. Today around 60 million tons of fish, mussels, crab, and other seagoing natural entities are cultivated throughout the planet every year. This is practically equivalent to the measure of sea fish and fish caught in the wild, which totaled 78.9 million tons in 2011. Asia, especially China, is the main aquaculture area, providing 89% of worldwide production (Mazur & Curtis, 2008). Aquaculture will keep on developing unequivocally and consequently make a huge commitment to furnishing the worldwide population with significant protein. A benefit of aquaculture is that many fewer feedstuffs are expected to cultivate fish and seafood than cows and pigs. It accepts 15 fold the amount of feed to deliver 1 kilogram of hamburger to create 1 kilogram of carp.

Aquaculture is subsequently a proficient asset technique as such of creating protein-rich food from animals. Current examinations researching likely advancements to 2050 demonstrate that aquaculture is fit for fulfilling the total population's developing requirement for fish. This continuous development, nonetheless, should not come at the expense of the climate or the environment. It is hazardous that aquaculture requires much wild fish prepared into fishmeal and fish oil and utilized as feed. Albeit the volumes of these products have been deteriorating for quite a long time, sometimes they utilize fish stocks that are not overseen economically. Aquaculture would thus be able to, in any case, be a supporter of the issue of over-fishing.

Endeavors are presently being made to diminish the measure of fishmeal and fish oil utilized in fish cultivating, not least since costs have taken off because of appeal in China. Many assessment groups are creating various sorts of greasy, protein-rich feed from potatoes and rapeseed. By and large, aquaculture production is as yet not economical. Functions require much energy and create supplement-rich flows, regularly directed into waterways and coastal waters in an untreated state. At that point, the waters become over-treated, causing algal blossom and oxygen-denied no man's lands. Researchers are presently creating strategies to break down the full life pattern of aquaculture offices – life cycle evaluations.

For quite a while, the business has been trying items for their natural similarity, accepting all angles from extracting the crude materials through to reusing. The intense raising of carp and shrimps in lakes is considered extremely destructive. It scores incredibly ineffectively in the existence cycle evaluation since it utilizes much to feed, produces supplement-rich profluent, and consumes much energy. However, lately, natural mindfulness has additionally gotten more broad in the aquaculture area. A difference in reasoning is apparent, especially in the industrialized countries that import enormous volumes of fish from Asia. Progressively dealers and clients are requesting merchandise that agree to natural guidelines. For certain years there have been manageability endorsements for marine catch fisheries, and items bearing this name are especially popular. Before long, the "Aquaculture Stewardship Council" endorsement, a seal of endorsement for aquaculture tasks, will show up in European business sectors. Merchants and producers wishing to have this testament should embrace to secure species, the climate, and the water in cultivated regions and follow an undeniable degree of social norms. We have a chance to outfit the further development of aquaculture towards maintainability. Current natural issues, over-fishing, and environmental change make this a question of direness.

10.7 CONCLUSION

Industrial aquaculture was designed during the 1970s. It is unique concerning conventional aquaculture that has been carried on for centuries. It started like the triumph of the Wild West, with incredible enthusiasm and the longing to get rich rapidly through "fish gold." No one at the time believed that it could have ecological outcomes. The area of plants, stock density, and taking care of productivity are altogether factors that must progressively be considered to lessen fish cultivating on coastal biological systems (Maulu et al., 2019).

The fish feed should consistently be overseen and conveyed productively by systems that guarantee admittance to benefit from interest, decreasing waste and bringing down change factors, clearly likewise to the upside of farmers. Buyers ought to be instructed to incline toward little pelagic fish, "bluefish," and herbivores to carnivores. In the last case, fishmeal ought to be supplanted with plant feed quite far, or more, all fishmeal ought to be acquired from results of fish handling. Land-based plants ought to have systems for bringing down supplement loads, like sedimentation lakes and essential and optional production dependent on the energy heap of wastewater. Marine plants ought to be adequately a long way from the coast to not influence coast ocean grass glades, even by supplements borne on ocean flows. Logical reviews like the Journal of Aquaculture Research and Development can play a determinant job in the discussion and the flow of thoughts and potential arrangements through open access distributing.

Through aquaculture, just a little part of the sea's region could fulfill the whole world's need for fish. This outcome is from an examination led by researchers from UC Santa Barbara, the Nature Conservancy, UCLA, and NOAA, the National Oceanic and Atmospheric Administration (detailed by Science Daily). That is a significant renowned group of specialists, and they propose that aquaculture, for sure we ordinarily call fish cultivating, can meet the developing issue of worldwide food insecurity.

References

1. Asche, F., Guttormsen, A. G., & Nielsen, R. (2013). Future challenges for the maturing Norwegian salmon aquaculture industry: An analysis of total factor productivity change from 1996 to 2008. *Aquaculture*, *396*, 43-50.
2. Allen, P. J., & Steeby, J. A. (2012). Aquaculture: challenges and promise. *Nature Education*, *2*(11), 12.
3. Allsopp, M., Santillo, D., & Dorey, C. (2013). Sustainability in Aquaculture: Present Problems and Sustainable Solutions. *Ocean Yearbook*, *27*(1), 1-22.
4. Amenyogbe, E., Chen, G., Wang, Z., Lin, M., Lu, X., Atujona, D., & Abarike, E. D. (2018). A review of Ghana's aquaculture industry. *J Aquac Res Development*, *9*(545), 2.
5. Bentzon-Tilia, M., Sonnenschein, E. C., & Gram, L. (2016). Monitoring and managing microbes in aquaculture–Towards a sustainable industry. *Microbial biotechnology*, *9*(5), 576-584.
6. Bottema, M. J., Bush, S. R., & Oosterveer, P. (2021). Assuring aquaculture sustainability beyond the farm. *Marine Policy*, *132*, 104658.
7. Bjørndal, T. (2002). The competitiveness of the Chilean salmon aquaculture industry. *Aquaculture Economics & Management*, *6*(1-2), 97-116.
8. Bjelland, H. V., Føre, M., Lader, P., Kristiansen, D., Holmen, I. M., Fredheim, A., ... & Schjølberg, I. (2015, October). Exposed aquaculture in Norway. In *OCEANS 2015-MTS/IEEE Washington* (pp. 1-10). IEEE.

9. Colsoul, B., Boudry, P., Pérez-Parallé, M. L., Bratoš Cetinić, A., Hugh-Jones, T., Arzul, I., ... & Pogoda, B. (2021). Sustainable large-scale production of European flat oyster (Ostrea edulis) seed for ecological restoration and aquaculture: a review. *Reviews in Aquaculture*.
10. Costa-Pierce, B. A., & Chopin, T. (2021). The Hype, Fantasies and Realities of Aquaculture Development Globally and In Its New Geographies. *WORLD AQUACULTURE*, 23.
11. Cole, D. W., Cole, R., Gaydos, S. J., Gray, J., Hyland, G., Jacques, M. L., ... & Au, W. W. (2009). Aquaculture: Environmental, toxicological, and health issues. *International journal of hygiene and environmental health*, *212*(4), 369-377.
12. Chavanne, H., Janssen, K., Hofherr, J., Contini, F., Haffray, P., Komen, H., ... & Bargelloni, L. (2016). A comprehensive survey on selective breeding programs and seed market in the European aquaculture fish industry. *Aquaculture International*, *24*(5), 1287-1307.
13. Cressey, D. (2009). Aquaculture: future fish. *Nature News*, *458*(7237), 398-400.
14. Diana, J. S. (2012). Is lower intensity aquaculture a valuable means of producing food? An evaluation of its effects on near-shore and inland waters. *Reviews in Aquaculture*, *4*(4), 234-245.
15. Diana, J. S. (2009). Aquaculture production and biodiversity conservation. *Bioscience*, *59*(1), 27-38.
16. Diana, J. S., Egna, H. S., Chopin, T., Peterson, M. S., Cao, L., Pomeroy, R., ... & Cabello, F. (2013). Responsible aquaculture in 2050: valuing local conditions and human innovations will be key to success. *BioScience*, *63*(4), 255-262.
17. Díaz, P. A., Álvarez, G., Varela, D., Pérez-Santos, I., Díaz, M., Molinet, C., ... & Figueroa, R. I. (2019). Impacts of harmful algal blooms on the aquaculture industry: Chile as a case study. *Perspect. Phycol*, *6*(1-2), 39-50.
18. Edwards, P. (2015). Aquaculture environment interactions: past, present and likely future trends. *Aquaculture*, *447*, 2-14.
19. Ellingsen, H., Olaussen, J. O., & Utne, I. B. (2009). Environmental analysis of the Norwegian fishery and aquaculture industry—A

preliminary study focusing on farmed salmon. *Marine Policy*, *33*(3), 479-488.

20. Evers, N. (2010). Factors influencing the internationalisation of new ventures in the Irish aquaculture industry: An exploratory study. *Journal of International Entrepreneurship*, *8*(4), 392-416.
21. Erondu, E. S., & Anyanwu, P. E. (2005). Potential hazards and risks associated with the aquaculture industry. *African Journal of Biotechnology*, *4*(13).
22. Fiorella, K. J., Okronipa, H., Baker, K., & Heilpern, S. (2021). Contemporary aquaculture: implications for human nutrition. *Current Opinion in Biotechnology*, *70*, 83-90.
23. Føre, M., Frank, K., Norton, T., Svendsen, E., Alfredsen, J. A., Dempster, T., ... & Berckmans, D. (2018). Precision fish farming: a new framework to improve production in aquaculture. *biosystems engineering*, *173*, 176-193.
24. Frankic, A., & Hershner, C. (2003). Sustainable aquaculture: developing the promise of aquaculture. *Aquaculture International*, *11*(6), 517-530.
25. Gopakumar, K. (2003). Indian aquaculture. *Journal of Applied Aquaculture*, *13*(1-2), 1-10.
26. Gjedrem, T. (2012). Genetic improvement for the development of efficient global aquaculture: a personal opinion review. *Aquaculture*, *344*, 12-22.
27. Hall, S. G., Campbell, M. D., Campbell, V. M., Geddie, A., Frinsko, M. O., Greensword, M., ... & Young, S. (2021). Smart Systems to Enhance Sustainability and Add Value to Marine Aquaculture. In *2021 ASABE Annual International Virtual Meeting* (p. 1). American Society of Agricultural and Biological Engineers.
28. Holmen, I. M., Utne, I. B., & Haugen, S. (2018). Risk assessments in the Norwegian aquaculture industry: status and improved practice. *Aquacultural Engineering*, *83*, 65-75.
29. Jensen, F., Nielsen, M., & Nielsen, R. (2014). Increased competition for aquaculture from fisheries: Does improved fisheries management limit aquaculture growth?. *Fisheries Research*, *159*, 25-33.

30. Jothiswaran, V. V., Velumani, T., & Jayaraman, R. (2020). Application of artificial intelligence in fisheries and aquaculture. *Biotica Research Today*, *2*(6), 499-502.
31. Kaminski, A. M., Genschick, S., Kefi, A. S., & Kruijssen, F. (2018). Commercialization and upgrading in the aquaculture value chain in Zambia. *Aquaculture*, *493*, 355-364.
32. Lafont, M., Dupont, S., Cousin, P., Vallauri, A., & Dupont, C. (2019, June). Back to the future: IoT to improve aquaculture: Real-time monitoring and algorithmic prediction of water parameters for aquaculture needs. In *2019 Global IoT Summit (GIoTS)* (pp. 1-6). IEEE.
33. Liu, G., Sun, J., & Liu, S. (2015). From fisheries toward aquaculture. In *Developments in Aquaculture and Fisheries Science* (Vol. 39, pp. 25-35). Elsevier.
34. Llorente, I., & Luna, L. (2016). Bioeconomic modelling in aquaculture: an overview of the literature. *Aquaculture International*, *24*(4), 931-948.
35. Maulu, S., Munganga, B. P., Hasimuna, O. J., Haambiya, L. H., & Seemani, B. (2019). A review of the science and technology developments in Zambia's aquaculture industry. *J Aquac Res Development*, *10*(567), 2.
36. Mazur, N. A., & Curtis, A. L. (2008). Understanding community perceptions of aquaculture: lessons from Australia. *Aquaculture International*, *16*(6), 601-621.
37. Milstein, A. (2005). Polyculture in aquaculture. In *Animal Breeding Abstracts* (Vol. 73, No. 12). CABI Publishing.
38. Naylor, R. L., Hardy, R. W., Buschmann, A. H., Bush, S. R., Cao, L., Klinger, D. H., ... & Troell, M. (2021). A 20-year retrospective review of global aquaculture. *Nature*, *591*(7851), 551-563.
39. Rosas, V. T., Poersch, L. H., Romano, L. A., & Tesser, M. B. (2019). Feasibility of the use of Spirulina in aquaculture diets. *Reviews in Aquaculture*, *11*(4), 1367-1378.
40. Sandersen, H. T., & Kvalvik, I. (2015). Access to aquaculture sites: A wicked problem in Norwegian aquaculture development. *Maritime Studies*, *14*(1), 1-17.

41. Shah, M. R., Lutzu, G. A., Alam, A., Sarker, P., Chowdhury, M. K., Parsaeimehr, A., ... & Daroch, M. (2018). Microalgae in aquafeeds for a sustainable aquaculture industry. *Journal of Applied Phycology*, *30*(1), 197-213.
42. Shaalan, M., El-Mahdy, M., Saleh, M., & El-Matbouli, M. (2018). Aquaculture in Egypt: insights on the current trends and future perspectives for sustainable development. *Reviews in Fisheries Science & Aquaculture*, *26*(1), 99-110.
43. Skerritt, D. J., Arthur, R., Ebrahim, N., Le Brenne, V., Le Manach, F., Schuhbauer, A., ... & Sumaila, U. R. (2020). A 20-year retrospective on the provision of fisheries subsidies in the European Union. *ICES Journal of Marine Science*, *77*(7-8), 2741-2752.
44. Sorgeloos, P. (2013). Aquaculture: the blue biotechnology of the future. *World Aquac*, *44*, 16-25.
45. Stenton-Dozey, J. M., Heath, P., Ren, J. S., & Zamora, L. N. (2021). New Zealand aquaculture industry: research, opportunities and constraints for integrative multitrophic farming. *New Zealand Journal of Marine and Freshwater Research*, *55*(2), 265-285.
46. Sirakov, I., Velichkova, K., Stoyanova, S., & Staykov, Y. (2015). The importance of microalgae for aquaculture industry. Review. *International Journal of Fisheries and Aquatic Studies*, *2*(4), 81-84.
47. Teletchea, F., & Fontaine, P. (2014). Levels of domestication in fish: implications for the sustainable future of aquaculture. *Fish and fisheries*, *15*(2), 181-195.
48. Troell, M., Naylor, R. L., Metian, M., Beveridge, M., Tyedmers, P. H., Folke, C., ... & De Zeeuw, A. (2014). Does aquaculture add resilience to the global food system?. *Proceedings of the National Academy of Sciences*, *111*(37), 13257-13263.
49. Turcios, A. E., & Papenbrock, J. (2014). Sustainable treatment of aquaculture effluents—what can we learn from the past for the future?. *Sustainability*, *6*(2), 836-856.
50. Van Doan, H., Hoseinifar, S. H., Ringø, E., Ángeles Esteban, M., Dadar, M., Dawood, M. A., & Faggio, C. (2020). Host-associated probiotics: a key factor in sustainable aquaculture. *Reviews in fisheries science & aquaculture*, *28*(1), 16-42.

51. Verdegem, M. C. J., & Bosma, R. H. (2009). Water withdrawal for brackish and inland aquaculture, and options to produce more fish in ponds with present water use. *Water Policy*, *11*(S1), 52-68.
52. Wurts, W. A. (2000). Sustainable aquaculture in the twenty-first century. *Reviews in Fisheries Science*, *8*(2), 141-150.
53. Yang, H., Sturmer, L. N., & Baker, S. M. (2020). Molluscan shellfish aquaculture and production. *EDIS*, *2016*(3), 8-8.
54. Yang, X., Ramezani, R., Utne, I. B., Mosleh, A., & Lader, P. F. (2020). Operational limits for aquaculture operations from a risk and safety perspective. *Reliability Engineering & System Safety*, *204*, 107208.
55. Zhang, Y., Bleeker, A., & Liu, J. (2015). Nutrient discharge from China's aquaculture industry and associated environmental impacts. *Environmental Research Letters*, *10*(4), 045002.

INDEX